Big Data for Urban Sustainability

Stephen Jia Wang • Patrick Moriarty

Big Data for Urban Sustainability

A Human-Centered Perspective

 Springer

Stephen Jia Wang
School of Design
Royal College of Art
Kensington Gore
London, SW7 2EU, UK

Patrick Moriarty
Monash University-Caulfield Campus
Department of Design
Caulfield East, VIC, Australia

ISBN 978-3-319-73608-2   ISBN 978-3-319-73610-5   (eBook)
https://doi.org/10.1007/978-3-319-73610-5

Library of Congress Control Number: 2018934414

Printed on acid-free paper

This Springer imprint is published by the registered company Springer International Publishing AG part of Springer Nature.
The registered company address is: Gewerbestrasse 11, 6330 Cham, Switzerland

Preface

Today, the majority of the world's population live in cities, in contrast to 1950, when under 30% were urban residents. However, has human society created some kind of monster that has already started harming humankind and other creatures on this planet? If so, it's time to ask: where are we heading? Vast megacities endlessly expanding in both vertical and horizontal directions? More dense living environments? A more conflictual relationship between urban environment and the natural environment? Urban sustainability has become the key to the future of our civilization.

Due to the dense concentration of both population and industrial activities, the quality of life of city dwellers already faces various challenges, typically air and noise pollution, traffic congestion and social stress, to a much greater extent than their non-urban counterparts. Nevertheless, urban areas are usually wealthier than non-urban regions, so that roughly 70% of global greenhouse gas emissions, and a similar share of global energy and mineral resource use, can be ascribed to urban residents. However, the reality is that the urban population share will surely keep rising for some time, as projected by the UN. When this occurs, their share of global greenhouse gas emissions and resource use can likewise be expected to grow, especially in the large industrialising countries with huge populations, industrial output and economies, such as China and India. The vast impacts through the process of urbanisation will inevitably influence more people's lives with a wide range of implications, especially for lifestyle, health and well-being, environmental and social changes.

The concept of *Urban Sustainability* in this book has two main aspects: the physical and the social. Physical urban sustainability, in turn, has two components: first, local urban environmental problems, such as air and noise pollution, and second, environmental and resource problems such as air and noise pollution, and second, the environment and resource problems generated worldwide by urban production and consumption. Although urban sustainability is often restricted to biophysical sustainability—air and water pollution, global warming, resource depletion etc.—this book assumes a broader use of the term and includes the important question of liveability and social sustainability. Urban social sustainability takes the human *liveability* perspective, examining such issues as food, water, social security, equity, education and even the emotional states of residents.

Does living in cities generate additional health and well-being problems compared with their non-urban counterparts? Well, rapid urban expansions are almost always associated with negative impacts on the social interconnectedness of the city, often leading to a lack of equity for urban residents, particularly in access to urban infrastructure and other basic services, as well as income inequality. Since China has just experienced the world fastest urban expansion in the past 30 years, the book includes a first-hand investigation of urban dweller experiences in Zhuhai, a typical example of a growing Chinese city. Another important feature of this book is its coverage of the cities of both the OECD and lower and middle-income economies. Most of the growth in the global urban population is now occurring in Asia and Africa, largely driven by migration from rural areas to cities.

Beyond mere analyses of urban sustainability, this book also suggests and discusses possible solutions through the implementation of Big Data. Big Data is being increasingly advocated as a fresh and promising approach to urban challenges, particularly through the notions of 'smart cities' and the 'internet of things'. This book selects three crucial urban problems—energy use and transport (and their consequent greenhouse gas and air pollution emissions), and *health and well-being*—and critically examines the potential role of big data in providing better services, lowering costs or reducing the environmental impact of these sectors. Since the application of big data in these areas is only beginning, unlike in business (particularly retail) and scientific data analysis, the emphasis is on their *future* potential in the three selected areas.

To make the most out of this book, the reader may choose to start with the particular chapter which discusses the topic of greatest interest, then tailor the order of chapters according to your needs instead of following the listed order in the book. Chapters 1 and 2 introduce the many challenges facing urban sustainability. Chapter 1 deals with the resource and environmental problems arising from energy use and urban transport, with Chap. 2 examining health and well-being problems. Chapter 3 introduces the concepts of big data, the internet of things and smart cities in the context of the expanding data needs for cities and gives examples of their present implementation. Chapter 4 sounds a cautionary note about big data applications, including the need to adequately address privacy and reliability concerns, among others. Chapters 5, 6 and 7 examine successively the potential role for big data in urban energy reductions, sustainable urban transport, and improved urban health and well-being. Finally, Chap. 8 looks to the future (the year 2050) and, assuming that adequate responses are made to the challenges raised in Chap. 4, considers how big data could contribute to urban sustainability in a carbon-constrained world.

London, SW7 2EU, UK   Stephen Jia Wang
Caulfield East, VIC, Australia   Patrick Moriarty

Acknowledgements

Stephen Jia Wang

I would like to thank my wife, Hiromi Wang, who has taken care of my health during many difficult periods. Also, I want to thank my parents for their spiritual guidance through this journey. I should not forget to mention Amelia Y. Wang and Edward Y. Wang, my most beloved daughter and son, who have endured my busy schedule but still returned my smile whenever I look at them. I also want to express my appreciation to my student Yang, Chulin and her family for their tremendous efforts in assisting gathering the first-hand information from the South China region. Last but not least, I want to thank Dr. Patrick Moriarty, who has not only contributed as the co-author of this book but also provided significant guidance as a mentor. This book could not have been achieved without all their great support!

Patrick Moriarty

I would like to thank the Department of Design at Monash University for providing me with accommodation during the research for, and writing of, this book. I would like to also thank Geraldine for her support during the project.

Abbreviations

App	Application
AV	Automated vehicle
b-a-u	Business as usual
BD	Big data
BECCS	Bioenergy carbon capture and storage
BRT	Bus Rapid Transit
CBD	Central business district
CCS	Carbon capture and sequestration
CCTV	Closed-circuit TV
CDC	Centers for Disease Control (US)
CHP	Combined heat and power
CNG	Compressed natural gas
CO_2	Carbon dioxide
CO_2-eq	Carbon dioxide equivalent
EC	European Commission
EF	Ecological Footprint
EIA	Energy Information Administration (US)
EJ	Exajoule (10^{18} joule)
EU	European Union
FCD	Floating Car Data
GDP	Gross Domestic Product
GNP	Gross National Product
GHG	Greenhouse gas
GIS	Graphical information system
GJ	Gigajoule (10^9 joule)
GL	Gigalitre (10^9 litre)
GNI	Gross National Income
GPS	Global positioning system
Gt	Gigatonne (10^9 tonne)
GtC	Gigatonne carbon
GW	Gigawatt (10^9 watt)

HDI	Human Development Index
HIV	Human immunodeficiency virus
IAEA	International Atomic Energy Agency
ICLEI	International Council for Local Environmental Initiatives
ICT	Information and communication technology
IAEA	International Atomic Energy Agency
IEA	International Energy Agency
IHVS	Intelligent Highway Vehicle Systems
IoT	Internet of Things
IPCC	Intergovernmental Panel on Climate Change
IT	Information Technology
kW	Kilowatt (10^3 watt)
kWh	Kilowatt-hour
LBP	Low back pain
LBS	Location-Based Service
LCA	Life cycle analysis
Mboe	Million barrels of oil equivalent
μg m^{-3}	Microgram per cubic metre
MJ	Megajoule (10^6 joule)
Mt	Megatonne (10^6 tonne)
Mtoe	Million tonnes of oil equivalent
MW	Megawatt (10^6 watt)
MWh	Megawatt-hour
NHS	National Health System (UK)
OECD	Organization for Economic Cooperation and Development
OPEC	Organization of the Petroleum Exporting Countries
PEB	Pro-environmental behaviour
PES	Pervasive Environment Simulator
PM	Particulate matter
$PM_{2.5}$	Particulate matter with diameters <2.5 micrometres
POI	Point of Interest
PPP	Purchasing power parity
PTA	Personal Travel Assistant
PV	Photovoltaic
QOL	Quality of life
QS	Quantified Self
RCP	Representative Concentration Pathway
RE	Renewable energy
RFID	Radio Frequency Identifier
SARS	Sudden Acute Respiratory Syndrome
SO_2	Sulphur dioxide
SRM	Solar radiation management
TB	Terabytes (10^{12} bytes)
tpk	Trillion passenger-km
TWh	Terawatt-hr (10^{12} watt-hr)

UHI	Urban heat island
UK	United Kingdom
UN	United Nations
UNDP	United Nations Development Program
UTES	Urban Transport Energy Saver
V2G	Vehicle-to-grid
WEC	World Energy Council
WHO	World Health Organization

Contents

1 **The Urgent Need for Advancing Urban Sustainability** 1
   1.1   Introduction 1
   1.2   Global Environmental Sustainability: Challenges
       and Potential Solutions 3
   1.3   Urban Environmental Sustainability: Challenges
       and Potential Solutions 8
       1.3.1   General Considerations 9
       1.3.2   Eco-efficiency 12
       1.3.3   Urban Transport 12
       1.3.4   Urban Buildings and Household Energy Use 14
   1.4   Concluding Remarks and Summary of the Book 16
   References 18

2 **Urban Health and Well-Being Challenges** 23
   2.1   Introduction: Global Health and Well-Being Challenges 23
       2.1.1   Physical Health and Mortality Worldwide 23
       2.1.2   Well-Being: Another Component of Health 25
   2.2   Urban Health in OECD Cities 26
       2.2.1   Urban Air and Noise Pollution 27
       2.2.2   Climate Change Effects on Urban Health 28
       2.2.3   Stress and Mental Illness 29
       2.2.4   Discussion 30
   2.3   Urban Health and Liveability in Non-OECD Cities 30
       2.3.1   Urban Air Pollution and Climate Change Effects 31
       2.3.2   Mental Health, Well-being and Liveability 33
   2.4   Urban Health and Liveability in Chinese Cities 34
       2.4.1   Rapid Urbanisation in China 34
       2.4.2   Air Pollution in Chinese Cities 36
       2.4.3   Liveability in Chinese Cities 37

2.4.4   Zhuhai City Experiences 38
2.5   Discussion 39
References.. 40

3   **The Potential for Big Data for Urban Sustainability** 45
3.1   Introduction: Traditional Urban Data Collection............... 45
3.2   Sustainable Cities Will Need a Rising Volume of Data.......... 46
3.3   Big Data, the Internet of Things, and Smart Cities 48
3.4   Present Applications of Big Data/Internet of Things in Cities 51
3.4.1   Structural Health of Buildings and Other
Urban Infrastructure................................ 51
3.4.2   Waste Management 52
3.4.3   Air and Noise Pollution 52
3.4.4   Traffic Congestion/Management, and Parking 53
3.4.5   Public Transport Information and Promotion 54
3.4.6   Pedestrian Traffic Counts.......................... 54
3.4.7   Urban Energy Consumption 54
3.4.8   Urban Health and Well-Being 55
3.4.9   Urban Governance 55
3.5   Future Potential for Big Data in Cities 56
3.6   Discussion 59
References.. 61

4   **Barriers to the Implementation of Big Data**..................... 65
4.1   Introduction 65
4.2   Privacy Problems 65
4.3   Security Problems 69
4.4   Reliability Problems.................................... 70
4.5   Technical problems 73
4.6   Cost Problems 74
4.7   Individual and Institutional Resistance to Big Data Solutions 75
4.8   Discussion 76
References.. 77

5   **Big Data for Sustainable Urban Transport** 81
5.1   Introduction 81
5.1.1   The Portrait of a City from a Transport Viewpoint 81
5.1.2   General Approaches to Sustainable Transport........... 83
5.2   Shifting to More Energy Efficient Modes 85
5.2.1   Public Transport.................................. 85
5.2.2   Non-motorised Transport........................... 87
5.3   Reducing the Demand for Urban Transport................... 88
5.3.1   Reducing Travel with IT 89
5.3.2   Reducing Freight Transport......................... 91

5.4   Raising the Energy and GHG Efficiencies of Urban Transport. . . .   92
     5.4.1   Automated Vehicles: A Possible Solution
            for Improving Efficiency?   92
     5.4.2   Improving Vehicle Efficiency   94
5.5   Beyond Energy Efficiency: Traveller Well-Being and Comfort . . .   94
5.6   Case Study of a Personal Travel Assistant for Beijing.   95
     5.6.1   Background   95
     5.6.2   Description of the Application.   97
5.7   Discussion: Future Urban Travel 100
References. 101

6   **Big Data for Urban Energy Reductions** 105
6.1   Introduction 105
6.2   Smart Grids: A Necessary Part of Sustainable Energy 106
6.3   Urban Domestic Energy Consumption 108
6.4   Smart Grids and Electric Vehicle Charging 111
6.5   Smart Buildings 112
6.6   An Integrated View of Urban Energy Use 113
6.7   Discussion: Energy and Urban Sustainability 115
References. 116

7   **Big Data for Urban Health and Well-Being** 119
7.1   Introduction 119
7.2   The (Contested) Potential for Big Data in OECD Healthcare 120
     7.2.1   Examples of Big Data Applications in OECD
            Healthcare and Well-Being 122
     7.2.2   Taking Charge: The Quantified Self Movement
            and Online Self-Help Groups. 124
     7.2.3   Discussion 125
7.3   Big Data Applications in Non-OECD Healthcare
     and Well-Being 126
     7.3.1   Examples of Existing Big Data Applications
            in Non-OECD Healthcare 126
     7.3.2   The Role of Big Data in the 2014 West African
            Ebola Outbreak 127
     7.3.3   Discussion 128
7.4   Case Study: Instrumented Chair for Health and Comfort 129
     7.4.1   Introduction 129
     7.4.2   Implementation of the Virtual Spine 131
     7.4.3   Discussion 134
7.5   Discussion: The Potential Benefits and Risks
     of Health Big Data. 136
References. 136

8 Big Data for a Future World 141
    8.1 Introduction: The World in 2050 141
        8.1.1 A Changing Planet 141
        8.1.2 Responses to a Changing Planet 143
    8.2 The Role of Big Data in Cities in 2050 147
        8.2.1 Big Data in OECD Cities 147
        8.2.2 Big Data in the Cities of Industrialising Countries 150
    8.3 Discussion 152
    References 154

Index 157

Chapter 1
The Urgent Need for Advancing Urban Sustainability

1.1 Introduction

The worlds first cities were established many millennia ago in the Middle East, China and the Indus Valley. Nevertheless, in 1900, only an estimated 16.4% of the global population of 1650 million lived in cities, although it was 46% in Western Europe, the then leading region for urbanisation [17]. By 2015, the global share had risen to 54% or around 3970 million out of a world population of 7350 million [78]. The UN expects this rapid urbanisation to continue, with the world being two-thirds urban by 2050, although others consider this proportion as unlikely to be reached [48, 69]. A rising share of this urban population lives in *megacities*—defined as cities with ten million or more residents [78]. Already in many OECD and Latin American countries, urbanisation is 80–90% [53], and so cannot be expected to rise much more. However, it is much lower in tropical Africa and many Asian countries, with rural migration to cities the main cause of the rapid urban growth in these regions. Most of the world's megacities are now in Asia: 17 out of the 30 ten million-plus cities in 2015, with only five being in Western Europe or the US [77].

For many researchers, this forecast continued growth is a concern [e.g. [66]], because many socio-economic and sustainability problems of cities appear to be a non-linear function of city size, even though the rationale for cities is economies of scale. Nathaniel Baum-Snow and Ronni Pavan [5] found that income inequality increases with city size in the US. Luis Bettencourt et al. [7] have produced a general theory of scaling in cities, and showed that crime rates and traffic congestion also rise with city size. Others have shown that in the US, the spread of HIV infection correlates well with urban population density [59], and the same is likely true for other contagious diseases. While the public may support the principle of urban sustainability, it can be argued that high-density development is too costly for individual quality of life. Very high levels of population density can cause housing tensions, traffic congestion, and environmental degradation—all components of what has been termed 'big city disease'.

Nevertheless, some researchers have argued that this continued urban growth is inevitable [37], and good for global environmental sustainability. Their argument is that per capita greenhouse gas (GHG) emissions in major cities such as London and New York are half or less of the national average for each country [4, 16].

However, it must be remembered that large metropolitan areas form a *system* [6, 41]. Travel needs, for example, will be less for residents closer to the city centre, but much greater for those living on the metropolitan region fringes (or even beyond), who may not even be included within the official city boundaries. These lower travel needs for inner city residents occur because many city-wide or even region-wide services (for example, state government offices, or major sports and entertainment venues) are only available at or near the city centre. The inner areas of most cities also normally have a job surplus. Such location means that inner city residents will have less travel to these destinations than others living further out. Also, urban sustainability in future will likely require urban residents to be more *self-provisioning* in energy, food, and in many cases, fresh water. Today's large, densely populated cities may be less suited both for this self-provisioning and also for the use of passive solar energy [45, 48].

The world's cities already produce most of the global Gross National Product (GNP), and are the site of consumption for most food, final energy, and materials. According to one estimate, cities account for 55%, 73% and 85% of GNP in low, middle, and high income nations respectively [68], values that will rise in low and middle income countries if these further urbanise as expected.

Urban *environmental* sustainability has two aspects. The first concerns the urban environment itself, such as urban temperatures, air and water pollutants, and urban ecology. The second aspect concerns environmental or resource problems that can be largely *ascribed* to cities, such as emissions of GHGs, including chlorofluorocarbons. Thus, urban residents are responsible for around 75% of global CO_2 emissions, although most of these emissions usually occur elsewhere, for example at fossil fuel power stations supplying electricity to urban homes, offices and factories [16]. Similarly, a full accounting for travel energy for urban residents must also include their surface and air travel beyond the city boundaries.

It follows that an ecologically sustainable planet is not possible without sustainable cities, even if the spatial extent of cities is only about 2% of the Earth's land surface [26]. Many city governments around the world realise this [28], and have taken the lead in attempts to reduce their greenhouse gas emissions, for example, those cities participating in the International Council for Local Environmental Initiatives (ICLEI) [33]. The rest of this introductory chapter will look at the *environmental* sustainability of cities, particularly their energy use, GHG emissions, and air pollution emissions.

However, urban sustainability, broadly considered, also has two aspects. This book goes beyond narrow definitions of 'sustainability'—often characterised by a focus on the environment and energy aspects, where people's quality of life is often overlooked, or at least placed in a secondary position. It devotes particular attention to *liveability* aspects of sustainability, which emphasises the living conditions and ever changing needs/lifestyles of individuals in an urban environment with

Fig. 1.1 Challenges to urban sustainability

information overload. Figure 1.1 shows the four challenges to urban sustainability in this broad sense, which must all be dealt with simultaneously.

1. Socio-cultural challenges (caused by globalisation and rapid urbanisation)
2. Environmental challenges (caused by global climate—and other environmental—changes)
3. Liveability challenges (caused by changing individual needs and lifestyles)
4. Technological challenges (caused by big data and ICT technology development).

The present section has briefly examined the first challenge, and the remainder of this chapter examines in detail the environmental challenges facing the world, including its cities. The need for healthy and liveable cities is taken up in Chap. 2, while the remainder of the book examines the role of big data in providing solutions.

1.2 Global Environmental Sustainability: Challenges and Potential Solutions

The two challenges particularly relevant today to societies continuing in a business-as-usual fashion are depletion of global mineral resources, particularly for oil, and global warming [20, 31, 43, 58]. It may be thought that the present (August 2017) low international oil prices are indicative of an oil surplus, of a global production capacity exceeding demand. However, several lines of evidence suggest 'peak oil' is still a near-to-medium term threat. First, it is generally agreed that production of *conventional* oil has already peaked [55]. Second, it follows that the world will in future increasingly need to rely on *non-conventional* sources (such as tar sands and shale oil, and oil from deep water or polar regions), if present production levels are to be maintained, let alone increased. However, these sources, while probably abundant [29], tend to have much higher economic, GHG, and environmental costs than conventional oil. The current low prices mean that investment in developing these expensive sources could be curtailed, with implications for future global oil production capacity.

It is also important to note that existing fields are losing production capacity at the annual rate estimated at three to four million barrels per day [56]. Hence, unless similar extra production capacity is found annually, overall production will decline. Jörg Schindler [71], in his survey of the future availability of all fossil fuels, argued that oil production has been on an undulating plateau for a decade or so, with no sharp peak. Nevertheless, he does see production levels falling sharply in a few years. Although cities tend to be leaders in alternatively fuelled vehicles, it is still the case that global transport in 2014 was 92% reliant on petroleum-based fuels, with very little improvement on the 94% level in 1973. Absolute levels of oil use in transport have steadily risen [34].

Other researchers have felt that we needn't worry too much about peak oil, because of the urgent need to address global warming—as evidenced by the December 2015 climate agreement in Paris—makes such concerns irrelevant. The Intergovernmental Panel on Climate Change (IPCC) in their *Synthesis Report* [30], has urged that global temperature rises above pre-industrial should be limited to 2 °C or even 1.5 °C to avoid serious anthropogenic climate change. Apart from the direct effects of higher temperatures, the world faces another environmental challenge from climate change: rising sea levels. The IPCC report [30] anticipated that global sea levels rise in the present century would likely be <1 m above 1986–2005 levels. However, James Hansen, a prominent climatologist, has argued on the basis of the paleoclimatic record, modern observations, and modelling that multi-metre sea level rise can be expected by the end of this century [25]. This rise, he and his colleagues argued, will occur *even if* global temperature rises above pre-industrial are limited to 2 °C. If they are correct, the changes needed for climate mitigation are far more drastic than is usually acknowledged, and limiting temperature rise to 1.5 °C becomes even more urgent.

Christophe McGlade and Paul Ekins [40] have stressed that one consequence of such stringent limits for climate change is that the world will have to leave most fossil fuels in the ground, including an estimated one-third of all oil reserves, 50% of natural gas and 80% of coal reserves. Globally, all fossil fuels in 2014 provided 467.2 EJ (86.3% of all commercial energy), compared with 24 EJ (4.4%) for nuclear power, and 50.1 EJ (9.3%) for all renewable energy sources (EJ = Exajoule = 10^{18} J) [9].

There are only a limited number of options [44] for making global energy supply and transport systems sustainable, including:

- A rapid shift to nuclear and/or renewable energy sources
- Mechanical carbon capture and sequestration (CCS) for remaining fossil fuels, and even for bioenergy fuels (BECCS), or alternatively, biological removal of CO_2 from the atmosphere by afforestation, among other methods

- Implementation of geoengineering, particularly solar radiation management (SRM)
- Reductions in the total primary energy cities and other regions use, either by significantly improved energy efficiency, or energy conservation—reducing the ownership, or at least use, of energy using appliances and vehicles.

Even the International Atomic Energy Agency (IAEA) did not expect nuclear energy to increase its present low share of total primary energy over the coming decades: they considered its share of global electricity production would rise little from 12.3% in 2011 over the current decade, but after 2020 will *fall* to anywhere between 5.0% and 12.2% by 2050 [32]. Similarly, the US Energy Information Administration (EIA) did not foresee any significant rise in nuclear power's share of the global electricity market out to 2040 [19]. Given the level of citizen opposition to nuclear power in many OECD countries, and that the reactor fleet is ageing, nuclear power cannot be expected to play a major role in future energy security or climate mitigation [23]. This conclusion is borne out by a recent analysis of the so-called 'nuclear renaissance': nuclear reactors are closing prematurely in many countries, and companies with nuclear expertise (such as Westinghouse Electric) are closing down [24].

Renewable energy (RE) includes a mix of established sources (geothermal and hydro, for example, which have been utilised for over a century) that are now only growing slowly [47], and new sources, particularly solar, which are growing rapidly, but from a small base. The largest source of RE globally is biomass, which the IEA [34] have estimated as accounting for 10.3% of total primary energy in 2014, most of it in the form of fuel wood (and animal dung) burnt at low efficiency in low-income countries. However, it is likely that much of this use is not environmentally sustainable in that it lowers soil fertility or reduces forest cover, and so should not be considered a renewable energy source.

Overall, it is unlikely that production of non-carbon sources of energy can grow fast enough to compensate for the deep cuts needed for fossil fuels. The global technical potential for sustainably-produced geothermal, hydro and biomass energy is probably too low to make a major contribution [45, 46]. The two RE sources with the greatest potential are wind and solar energy. However, they are both intermittent sources, necessitating energy storage at higher levels of market penetration [50, 62]. If RE has to supply all energy needs, not just electricity, both conversion of electrical energy to another energy carrier, such as methanol or hydrogen, will also be necessary. Conversion and storage of energy will not only be expensive, but will also entail substantial energy losses.

The future availability of some important non-energy minerals is also in doubt [22, 45]. Although recycling rates for many materials could be improved, for some minerals it is already high, and if present trends continue, growth in output will be needed as presently low income countries industrialise. What is particularly important is that the RE sources with the greatest technical potential—wind and solar—in contrast to the long-established RE sources biomass and hydropower, increasingly incorporate exotic materials in their production to improve their energy efficiency. These minerals are often of low abundance in the Earth's crust, so, as for conventional minerals, rising production will entail both rising energy costs and pollution for production as ore concentrations fall [22]. There is thus the risk of mere *problem shifting*—reducing CO_2 emissions at the expense of other environmental problems [49, 50, 79].

Carbon dioxide (CO_2) removal from the atmosphere can occur either by mechanical or biological means. Mechanical CO_2 removal, whether by CO_2 capture from the exhaust stacks of power plants or other large industrial sources like oil refineries or steel plants, or by CO_2 removal from ambient air, has both high energy and economic costs [45]. Removing CO_2 by biological means—encouraging carbon storage in soils, or afforestation/reforestation—is cheaper, but its potential may be limited, and may conflict with the essential aim of improving, or at least maintaining, biodiversity [21, 63]. Others have argued that both the global potential for biological sequestration is too small to make a decisive difference [72], and that for boreal afforestation, where green tree cover would replace snow cover, the climate forcing effects of decreased regional albedo (the albedo is the fraction of insolation reflected directly back into space, presently at a global average of about 0.3) will offset the beneficial effect of increased carbon storage in the new forests [35].

The latest IPCC mitigation report [29] placed much reliance on BECCS as a means of keeping the global temperature at safe levels. This largely unproven technology [2] is attractive because, unlike CCS which can only be applied to current emissions, it offers the possibility of drawing down atmospheric CO_2 levels, in the same way as air capture would, but at much lower energy costs. If widely implemented, it would allow global CO_2 emissions and atmospheric CO_2 levels to temporarily overshoot safe levels, thus buying time for other policies to be implemented. However, given the competition between bioenergy and other uses (agriculture, forestry, and other biomaterials) for vital resources such as suitable land and water, the global potential for sustainable bioenergy may be far too low for BECCS to make much difference [51]. Such competition can only grow more intense as the world population moves toward 11.2 billion, the upwardly revised median UN estimate for the year 2100 [78]. In any case, there are numerous problems for carbon sequestration, including citizen opposition, liability and other legal problems, risks of (and from) seismicity, and possible limitations on the volumes of CO_2 that can be safely and securely sequestered each year [45].

Another means by which fossil fuel use could continue unabated—assuming that reserves are in fact sufficient to allow this for decades to come—is to implement *geoengineering*. With geoengineering, the idea is to modify the environment, usually but not always the atmosphere, on a global, or at least regional scale. The most commonly proposed method is SRM, which would involve annually injecting millions of tonnes of aerosols (sulphates are most commonly discussed) into the lower stratosphere in order to cool the planet by increasing the Earth's albedo. The albedo at the regional level could also be enhanced by using reflective coatings on urban roads and roofs, covering vast desert areas with reflective sheets, or even changing the leaf albedo of crops [29].

SRM has several advantages. It appears to be much cheaper than conventional mitigation options such as increased use of RE, or implementing CCS [39]. Compared with conventional mitigation methods, it could be quickly implemented, and the resultant cooling would occur in less than a year. Also, given that the aerosols would rain out over a year or two, and so need continuous replacement, it could be quickly stopped if serious side effects were felt. On the other hand, *ocean acidification*, and with it the risk of destabilising ocean's ecosystems, would continue apace, depletion of fossil fuels would once more be on the agenda, and average global precipitation would likely decline, a problem in a world where many regions are already experiencing water shortages [45]. Further, since the benefits and environmental costs of geoengineering would be unevenly distributed among nations, it would be very difficult to garner international consensus for SRM: the net losers would be very reluctant to agree to an international SRM project.

In summary, it is improbable that a largely unproven and politically controversial technology such as CCS, even together with the output from non-carbon energy sources, can expand fast enough to reduce CO_2 emissions in the time frame available. SRM, assuming it does work as planned, will have serious side effects, some of which are already known, but likely others which are presently unknown. Probably for this reason, SRM was discussed, but not included, as a possible solution to climate change, in the latest IPCC report [30]. Instead, it is likely that the brunt of carbon reductions (if indeed they occur) will need to come from reductions in primary energy use, both from energy efficiency improvements and from energy conservation—reducing our use of energy-consuming devices.

Put simply, global consumption levels for energy are far too high, mainly the result of massive subsidies to fossil fuels, which in any case are a one-off bounty. The International Monetary Fund has calculated that fossil fuel subsidies totalled $5.3 *trillion* (about 6.5% of global Gross Domestic Product) for the year 2015 [11]. A minor part of this calculated subsidy is for consumer subsidies (i.e. international price minus consumer price). Most (over 80%) of the subsidy was for *negative externalities*, such as air pollution and GHG emissions from transport. Increasingly, RE will need to supply our energy needs, but not necessarily at today's heavily-subsidised level of primary energy use: the world will need to reduce energy consumption. How such reductions could be made in an urban context are discussed in the following section.

So how can the profound changes in global transport and energy use can be brought about? In climate science, scientists use the term *climate forcing* [29], measured in W/m^2, to quantify the effects of GHGs on global surface temperature. It is likely that an analogous external forcing will be needed to bring about the individual and political changes needed for future ecological sustainability. We suggest that a rapid rise in extreme climate events—heat waves, heavy rainfall and flooding, storms and cyclones—will provide the impetus for change, as more and more of the world's population directly experience climate change [48]. Such changes will be helped by the realisation that consumers do not desire energy *per se*; what they really want is *energy services*, in the same way as access, not mobility, is what transport provides [41].

Table 1.1 Energy and economic data for various countries, 2014

Country	GNI/capita (PPP[a] 2010 USD)	kWh/capita (electricity only)	kW/capita (all energy)	CO_2 emissions (tonne/capita)
Australia	44,158	10,002	7.04	15.81
China	12,547	3938	2.97	6.66
Eritrea	1334	63	0.21	0.11
Ethiopia	1402	70	0.66	0.09
Iceland	40,636	54,000	23.82	6.25
India	5329	805	0.85	1.56
Japan	34,905	7829	4.62	9.35
Norway	59,916	23,000	7.44	6.87
UK	37,793	5131	3.69	6.31
US	50,621	12,962	9.21	16.22
OECD	36,494	8028	5.52	9.36
World	10,053	3030	2.51	4.47

Source: [34]
[a]PPP = Purchase parity Pricing

1.3 Urban Environmental Sustainability: Challenges and Potential Solutions

So far we have considered resource depletion and climate change from a global viewpoint. We now look at the implications for cities, recognising the enormous differences between cities in the world, both in GNI per capita and in resource and energy consumption. Since comparable city-level data are seldom available, national data has been used. For example, at the national level, electricity consumption per capita in 2014 ranged from 54,000 kWh in Iceland down to 63 kWh in Eritrea (see Table 1.1). Many tropical African countries, including Eritrea, also have low national levels of access to electricity. In comparison with the global '2kW society' advocated by Daniel Spreng [73] as representing a continuous average power (based on total primary energy) per capita needed for adequate energy use, Iceland in 2014 averaged 23.82 kW while in Eritrea the figure was 0.21 kW. As Table 1.1 shows, the 2014 world figure of 2.51 kW already exceeds the proposed 2 kW limit. (But even 2 kW per capita may be too high, and of course if followed, total energy use would continue to grow along with world population.) It follows that for most urban residents in many low-income countries, especially in tropical Africa, energy use, including that for transport, will need to increase to achieve a decent standard of living. Conversely, energy and resource consumption in high-income cities, and even middle-income cities, will need to fall. The national figure for China already exceeds the 2 kW limit.

For many OECD countries overall, absolute levels of primary energy have steadily fallen in recent decades, as also have CO_2 emissions from energy and industrial sources. In the UK for example, energy-related CO_2 emissions peaked in 1973 at 718 Mt, but by 2015 had fallen to 437 Mt [9]. However much of this fall in both energy and emissions can be attributed to OECD de-industrialisation, and the corresponding rise of Asian manufacturing, particularly in China. When the net energy and CO_2 emissions embodied in national imports and exports are included, most or even all of this energy and emissions decline vanishes [15].

The most important uses of final energy in cities are for transport, buildings (which can be private residences, commercial buildings, or public buildings such as government schools or offices), and industry. This book will not consider energy efficiency improvements in industry, or in power stations. These topics are well-covered in the latest IPCC reports [29, 30]. However, it will discuss smart grids, which will be necessitated by increased reliance on intermittent RE (see Chap. 6).

1.3.1 General Considerations

One simplified but useful way of viewing cities, particularly from an environmental sustainability viewpoint, is as a system, with inputs and outputs. Cities everywhere must usually import fresh water, energy (gas, oil, electricity, etc.), food, building and other materials, as well as various products manufactured elsewhere. These inputs are used up and degraded, producing a variety of waste products—air pollution, sewage, solid wastes, waste heat and so on. People and goods circulate within and between cities, requiring massive transport infrastructures and producing congestion. Finally, vast (and rapidly rising) amounts of information also circulate within and between cities, an important consideration we will return to in later chapters.

Both energy, water, and road and rail transport infrastructure demand varies by time of day, the day of the week, and also seasonally. The resultant peaking in usage means that, for example, electric power plants and transmission/distribution line capacity must be sufficient to meet the peak demand, but is then under-utilised for the remainder of the time. Similarly for transport infrastructure, except that given the space limitations in most large cities, it is usually not possible to ever meet the desired peak demand for road space, resulting in traffic congestion. In Tokyo, for instance, the number of vehicular trips within 50 km of the central railway station has not increased at all over the past few decades, and has fallen for road travel, despite growth in the relevant urban population [74].

Energy use in cities, whether for transport or other uses, is the main producer of air pollution, which was recently estimated to contribute to the deaths of about 1.6 million persons each year in China alone [67]. Air pollutants can also cause damage to building exteriors and reduce crop production. The worst air pollution occurs in the megacities of Asia, much of it the result of the lower or poorly-enforced emission standards for the rapidly rising number of road vehicles. However, another important problem in low income cities is *indoor* air pollution, whose main source is inefficient cooking stoves. Research is now being undertaken to find ways of both improving their efficiency and reducing their pollutants. The rising urban temperatures expected in future could worsen the air pollution in all cities [48]. Fortunately, reducing fossil fuel use—in OECD cities at least—can reduce both some forms of urban air pollution and ascribed GHG emissions. This synergy is not the case, however, for sulphate aerosols, which to some extent reflect insolation directly back into space (thus functioning as an unintended form of SRM) and so reduce climate forcing from GHGs [45]. In this case, cleaning up the local urban environment will to some extent be at the expense of the global environment. The health effects of urban pollution are considered in detail in Chap. 2.

The waste heat produced from transport and other energy use is also an important contributing factor to the *Urban Heat Island* (UHI) effect. This UHI effect results in large cities often being several degree Celsius warmer than the surrounding countryside, with the effect being more pronounced for daytime and summer temperatures. (Only for arid climate cities such as Las Vegas will cities be cooler than the surrounding region, because of greater vegetation evapotranspiration.) Other factors contributing to UHI include a high proportion of impermeable surfaces which prevent the cooling effect of evapotranspiration, and tall buildings which block outward radiation to space [36, 48, 61]. UHI effects will exacerbate the intensity of heat waves, expected to occur more frequently under global climate change.

Many important cities are either located on sea coasts or major rivers, or even both, raising the risk of inundations from rising sea levels (even from lower estimates of future sea level rise) or flooding from higher intensity rains that further global warming is expected to bring. Coastal land subsidence caused by pumping water from coastal aquifers, or non-replenishment of silt because of damming of rivers, also heightens the risk of coastal flooding. Seven major cities, all in Asia (Dhaka, Manila, Bangkok, Yangon, Jakarta, Ho Chi Minh City, Kolkata) are considered to face high risk not only from rising sea levels but also from other natural disasters that climate change will exacerbate [57]. Nor will OECD cities be exempt: New Orleans is also at risk from rising sea levels, and superstorm Sandy in October 2012 caused an estimated \$30–\$50 billion damages (2012 prices) in coastal New York and New Jersey [48]. While it cannot be said that this superstorm was a direct result of climate change, its likelihood of occurrence was increased.

But resource depletion and climate change and their consequences are not the only environmental challenges cities face. Because of their dense concentrations of people and infrastructure, cities are more vulnerable to natural hazards than non-urban areas. However, like urban flooding, their frequency and severity will likely rise with ongoing climate change. The frequency of urban landslides can be expected to rise in hilly areas, the result of both higher rainfall intensity and increased settlement of unstable steep urban hillsides by swelling urban populations, particularly in industrialising countries, and even the process of urbanisation itself [85]. In addition to climate-related problems, many cities are already at risk from earthquakes, as evidenced by the destruction in Kathmandu in 2015. Even volcanic activity can be affected by climate change. A handful of cities is even located near active volcanoes. Mt Ranier, near Seattle, will lose its ice cap as warming progresses, which could trigger devastating landslides by destabilising the summit cone [65]. The UN [77] has even listed cities facing one or more of six environmental hazards: cyclones, droughts, earthquakes, floods, landslides, volcanoes. A number of cities, including some major ones, even today face three or more of these hazards.

The urban ecology of cities is also important [60, 76]. Humphries [27] has described the trend toward treating the city as an *ecosystem*. This idea develops further the description of cities as a system, as discussed at the beginning of this section. Urban green spaces and parks serve several important functions: they can be both aesthetic spaces and places for relaxation and recreation. They also provide a habitat for urban birds and other fauna. (Indeed, Los Angles is one of the world's biodiversity hotspots, and urban areas are regarded as generally important for biodiversity conservation [3]). But trees and other vegetation in urban areas are important in more practical ways. Evapotranspiration from vegetation can reduce the UHI effect and produce local cooling, and the shade that trees can provide can be an important part of passive solar energy. Some tree species can also absorb pollutants, either through filtering out particulate matter (PM) or gaseous absorption. But tree species vary in their effectiveness: in Chinese cities, one study found a 14-fold difference in their ability to remove airborne PM [83].

Nevertheless, it is important that such shading does not conflict with provision of energy from rooftop solar PV cells. Other possible downsides of trees in urban areas is that the foliage can reduce cross-ventilation, the roots can increase maintenance costs of roads and footpaths, and, in drier climates, the trees may require watering. Under northern climatic conditions, Finnish researchers have also argued that the ability of trees to remove urban air pollution is limited [70]. Some tree species will harm people with pollen allergies, others may have foliage which limits air circulation and thereby increase street level particulate matter [83]. In brief, urban areas will need many more trees, carefully located and with attention to species selection, but even then tradeoffs will have to be made.

Cities have two possible responses to the risks from climate change. They can help mitigate climate change by reducing GHG emissions, as we discuss in detail below for transport and building energy use. One difficulty, however, is that any benefits will accrue to the world overall, and only in a minor way to the city itself. With climate *adaptation* [10], in contrast, most benefits are retained by the city, which should help gain support for such policies. In fact, adaptation is already practiced in cities in response to emergencies such as flooding, storms, and disease outbreaks. In any case, given that the adverse effects of climate change are already being experienced, cities will need to adopt both mitigation and adaptation policies.

Nevertheless, care will need to be taken with climate adaptation, for several reasons. Cities must not regard adaptation as a substitute for urgent action on climate mitigation. Further, adaptation measures can sometimes conflict with mitigation policies, as is the case with the growing global use of air conditioning (see Sect. 1.3.3). Care must also be taken to ensure that adaptation does not worsen inequality within cities. For example reducing the risk of flooding from whatever cause in one area of the city can be at the expense of increased risk for those in other, unprotected areas [48]. Finally, if the world continues to do little about mitigating climate change—and since the time of the first IPCC report in 1991, atmospheric CO_2 concentrations have risen 55%, to over 400 ppm in 2016 [9]—the limits of adaptation as a local response will prove increasingly ineffective.

1.3.2   Eco-efficiency

Table 1.1 showed the huge disparity in per capita energy use between different countries. Section 1.2, on the other hand, discussed the global environmental and resource problems the world faces. Evidently, the consumption disparity cannot be resolved by merely increasing global primary energy consumption levels. What must be increased greatly is *eco-efficiency* [54], which can be broadly defined as maximising the useful services we obtain from a given level of consumption while minimising the impact on the environment.

Sections 1.3.3 and 1.3.4 below discuss how eco-efficiency can be raised for both urban transport and domestic energy consumption. Thinking in eco-efficiency terms requires that we look at traditional practices in a new way. For urban transport, we have to go beyond merely increasing the vehicular passenger-km from each unit of primary input energy, and question what travel itself is for. Can we arrange matters so that the urban residents as a whole can travel to and from work, school, shops, etc., with far less total vehicular travel? Similarly, for domestic energy use, particularly heating and cooling, we need to go beyond merely trying to minimise the energy needed to keep the interior of buildings at 20 °C. The actions of the building occupants can heavily influence the energy needed for thermal comfort, as discussed below.

1.3.3   Urban Transport

Reducing urban transport energy use not only reduces the level of GHGs that can be directly attributed to urban residents but can also help cut both urban air and noise pollution, which is of immediate health benefit to urban residents (see Chap. 2). In general, the environmental sustainability of urban transport, both passenger and freight, can be improved by adopting one or more of the following general approaches, which apply regardless of city location or income level [41]:

- Using alternative fuels and propulsion systems. This approach can often enable not only energy efficiency improvements but also reductions in both GHGs and local air pollution. Possibilities include alternative fuels in internal combustion engines, such as ethanol or methanol from biomass, hydrogen, or natural gas; hydrogen fuel cell vehicles; and electric drive vehicles—hybrid, plug-in, or full battery electric vehicles (EVs). In 2015, for example, natural gas vehicles numbered 22.74 million globally; nearly all of these were in non-OECD countries [82]. Electric drive, whether used in electric public transport, or in hybrid or battery cars, is increasingly seen as the best way of improving overall vehicle efficiency, since it both allows regenerative braking and eliminates the need for engine idling.
- Improving the energy efficiency of all transport vehicles, whether for passenger or freight transport. Two approaches are possible: improving the engine efficiency, or reducing the *road load*. The first attempts to raise the share of input

energy (from the fuel tank or overhead power lines) that turns the wheels, and is sometimes called the 'tank-to-wheels' efficiency [42]. All vehicles use the energy output from the drive shaft in overcoming the road load, which consists of the sum of rolling friction, air friction, and inertial resistance. These three loads can all be reduced by improved vehicle design. The most important road load in an urban context, inertial resistance from the stop-start conditions of urban traffic, can be lowered either by reducing vehicle mass, or, as mentioned above, by regenerative braking. Note that unlike many of the possible engine efficiency improvements, reducing the road load can be made independently of the fuel used or even the propulsion system. The scope for energy efficiency improvements are considered to be very large, but so far the many advances in efficiency have been offset by the energy demands of auxiliary services (power steering, entertainment, air-conditioning) and improved vehicle performance [42]. Also needed is to reduce the energy used to manufacture vehicles. Increasing the lifespan of vehicles spreads these embodied energy costs over more years, and so lowers the total annual energy costs of vehicles, which include energy costs for manufacture and maintenance, as well as those for operating vehicles (i.e. fuel energy).

- Increasing the load factor for freight vehicles, and the occupancy rate for all passenger vehicles. There is less scope for urban freight, mainly because on-time delivery is at least as important to the business sector as energy efficiency. For passenger travel in OECD cities, there is often a very large potential for improved loadings, especially for passenger cars and for off-peak public transport. In OECD countries, overall car occupancy rates are typically around 1.5 persons or even less, giving a 30% seat occupancy for a five seater vehicle [52]. At peak hours, car occupancy can fall to 1.1–1.2 persons, important since traffic congestion is then at its worst.

- Increasing the market share of energy-efficient transport modes, both for passenger and freight. Energy efficiency can be best measured as passenger-km or tonne-km per megajoule (MJ) of primary energy for passenger and freight respectively. Primary energy, rather than litres of fuel, is used to allow for comparisons between electric traction and petroleum-based fuels. As usually measured, it depends on loading rates: a high-occupancy car is more energy efficient than a nearly empty train. Nearly all urban freight is carried by road, and an increasing share by smaller (and thus less energy efficient) commercial vehicles. As explained above for load factors, energy efficiency for freight is usually seen by operators as less important than on-time delivery. For passenger travel, however, not only are the larger public transport vehicles more energy efficient on a seat-km per primary MJ basis, but the scope for higher occupancy rates is also greater. The reason is that the average occupancy rate for private vehicles is partly constrained by (and in fact correlates strongly with) average household size, which has been falling in OECD countries for decades [52]. Public transport is not so constrained and has the further advantage that occupancy rates >100% can occur on services when some passengers are standing—although passengers may see this as overcrowding. Of course, the most energy efficient modes of passenger travel are walking and cycling. These modes have great potential, even in low and medium density OECD cities, given that many trips are very short and hence suited to these modes.

Two important questions arise about these energy efficiency approaches. First, can these methods, singly or together, significantly reduce both transport energy use and related GHG emissions? Second, even if they can, can they do it in the limited time frame we have left for mitigating climate change? Previous research, as well as the experience of the past few decades, suggests that with the possible exception of modal shift in urban passenger transport, they can only be of minor use with anything like existing transport policies in place.

We will need to go further than these four approaches, and improve the efficiency of urban transport itself. Just as the four above approaches aim to increase passenger-km per MJ of primary energy, *transport efficiency* would try to increase the *access* to desired out-of-home activities that can be obtained from each passenger-km of travel. Improving transport efficiency need not lower the quality of urban life, since most travel is considered to be a *derived* demand; travellers endure the monetary and time costs of such travel to access destinations such as shops, schools, and workplaces. Recent decades have seen the *suburbanisation* of destinations such as retail centres and workplaces in many OECD cities. Such suburbanisation had the potential to lower urban travel levels, but paradoxically urban travel per capita rose in most OECD cities, probably because of the convenience of car travel [41]. The important point is that there is a great untapped potential for reducing urban travel, at least in the cities of OECD countries, and with it, the accompanying air and noise pollution, GHG emissions, traffic collisions and injuries, and travel costs.

Although in some OECD cities, a bare majority of households own one or more vehicles [1], car ownership is far lower for residents of most cities in the industrialising world. Unlike OECD cities, car ownership in these cities is higher in urban than in non-urban areas, but it is still often very low, mainly because ownership is too expensive for most households. On a national basis, vehicles per 1000 persons are still <10 in many African countries, compared with over 500 typical of OECD countries [81]. Even so, the injury and death rate per 1000 vehicles tends to be very high in low-vehicle ownership countries, mainly because of the very high share of unprotected road users (pedestrians, pedal cyclists, motor cyclists). In China, pedestrians and cyclists made up 60% of traffic deaths [18]. If non-motorised modes are to be both safer and more enjoyable to use, travel speeds will have to be reduced in urban areas, not only in industrialising countries, but in high-income countries as well. Only then can a major shift back to these modes occur.

1.3.4   Urban Buildings and Household Energy Use

Another important contributor to urban energy use is that used in buildings, whether for space heating and cooling, water heating, lighting and powering the numerous energy-using devices in homes, workplaces, and shops. In a similar manner to transport, energy reductions in the appliances in buildings can come about either through improvements in the efficiency of energy-using devices—such as lights, refrigerators or air conditioners—or from reduced use (or lower settings) for these devices. Although there have been great gains in the efficiency of lighting (in terms of lumen per watt), TV sets and other devices [45], global electrical energy use is still growing strongly, rising from about 15,000 TWh in the year 2000 to 24,800 TWh in 2016 [9]. Gains in large appliance efficiency have been more than offset by rising levels of ownership. This, in turn, has been driven by income growth, particularly among the urban middle classes in industrialising countries and by the decline in the real costs of purchase for these appliances [80]. Clearly, energy efficiency improvements alone are unlikely to deliver more than marginal global energy and GHG emissions reductions—the use of these appliances must be curtailed.

Energy is also consumed in constructing and maintaining buildings, as well as in the provision and operation of public infrastructure such as water supply or street lighting. Increasingly, life cycle analysis (LCA) is used to examine the energy costs of constructing buildings over their useful life, then demolishing them and removing waste at the end of their lives. The useful lifespan of buildings is much greater than for vehicles, and increasing it further can again lower annual total energy costs. The energy costs of building materials can also often be lowered by using timber instead of more energy-intensive steel and concrete. The timber can then often be reused as a building material, then finally combusted for energy, perhaps in combined heat and power systems [51].

Much research has looked at ways of reducing domestic energy consumption (e.g. [13]), that part of total urban energy use—along with private vehicle fuel use— over which households have direct control. Nearly all residential energy use in OECD cities presently comes from reticulated supplies of natural gas and electricity. Few levers are available (apart from hefty price rises) for reducing domestic energy use, in marked contrast to reducing private transport energy use. Transport authorities can (and do) regulate allowable vehicle emission levels, minimum fuel efficiency standards, and speed limits. They can restrict parking and road space availability, impose charging for road space, and so encourage alternatives to car travel. But apart from efficiency standards for new appliances and thermal standards for new buildings, authorities must largely rely on moral persuasion (or fuel price rises) to reduce domestic energy use. Such reductions will be made even more difficult by the continuing decline in average household size (which is a worldwide phenomenon) and ensuing increases in floor space per occupant, which tends to raise the energy costs of space heating or cooling on a per capita basis.

For buildings in all cities, there is an often untapped potential for more use of passive solar energy. Many vernacular forms of architecture rely on passive solar energy as a matter of necessity, but in high income cities, fossil-fuelled heating and cooling systems have dominated, with the number of households with air conditioning systems showing continued growth in traditional OECD countries, as well as strong growth in industrialising countries like Mexico. As global temperatures further increase, it is likely that ownership, and with it electricity use, will rise in a non-linear fashion [14]. Of course, this private adaptation to global warming will merely exacerbate global climate change, and is better viewed as a mal-adaption. Passive solar energy can be seen as a way of saving on heating/cooling and lighting energy. Also, households worldwide are now increasingly installing solar photovoltaic (PV) cells and solar hot water systems; these installations of active energy help reduce fossil fuel consumption. Particularly in tropical Africa, where in many countries <20% of households were connected to mains electricity in 2012 [84], PV cells and a storage battery enable householders to run one or two lights and a TV set in the evenings.

Households in urban areas can also reduce their ecological footprint by installing rainwater tanks, and by urban agriculture, either on private or publicly-owned plots. Particularly in the cities of low income countries, urban farming is very common, with an estimated 800 million urban residents worldwide growing food or keeping livestock. Once common in the cities of OECD countries, it is now enjoying a resurgence [38]. Urban farming can be merely a hobby of course, but it can also be an important way of ensuring food for low-income households. Local food production also helps eliminate the many 'food miles' of transport—and so the consequential energy use and GHG emissions—that most food presently consumed in high-income cities make, a result of refrigeration and cheap sea freight transport.

1.4   Concluding Remarks and Summary of the Book

Luis Bettencourt [8] regarded the growth of global urbanisation and the unprecedented rise and spread of information and communication technologies (ICTs) as the most important trends in the modern world. Another is the ever-changing needs and lifestyles of urban residents. To these three, we would add a fourth, which we regard as even more important than the three mentioned: the rise of global environmental challenges, as detailed by Steffen et al. [75]. This chapter has looked at the first trend (global urbanisation) in the light of the fourth trend, discussing the various environmental and resource problems that cities will increasingly face in the coming decades. Ongoing global warming is also expected to have a disproportionate impact on cities and the risks their residents will face, given their dense concentrations of both people and built infrastructure [48].

In the chapters that follow in this book, we try to show how the second trend toward pervasive use of ICTs, and specifically the vast quantities of data they generate, can be used to make cities more ecologically sustainable, both in the face of the increasing risks they face, and the need to dramatically reduce their GHG emissions [64]. Satellites now have the potential to track GHG emissions at the city level [12]. While acknowledging that big data also has many potential commercial benefits, some of which are already starting to be tapped, the focus in this book is on improving urban sustainability, from both a biophysical and socioeconomic point of view. John Day and Charles Hall [16], in their book *America's Most Sustainable Cities and Regions*, have argued that there is a need for a radical rethink in our search for sustainable urban solutions, given the urgent problems cities will face. This book attempts such a radical rethink.

Chapter 2 is a detailed look at urban sustainability from a health and well-being viewpoint, and is thus a complement to the present chapter, which emphasised the biophysical aspects of urban sustainability. Two globally important health problems are the ageing of the population and the widespread rise in health costs as a share of national income. The health and well-being of urban residents, which goes beyond the mere absence of physical and mental illness, are examined for both OECD and non-OECD countries. The creation of a truly sustainable city in the future not only requires simple increases in energy efficiency. The personal quality of life for urban residents—the creation of livable, stable and vibrant communities—is also important, and will become increasingly so in future. The urban problems of China, home to 20% of the global urban population, are given particular emphasis.

Chapter 3 first looks at existing data collection in cities, and its limitations, then at the reasons why making cities sustainable will need vastly increased amounts of data in future. It next describes the rise of the Internet of Things (IoT) and how the data from vast numbers of urban sensors could make cities 'smarter'. The chapter gives a number of examples of how big data and IoT is presently being used in various cities. Since the impact on sustainability in smart cities is presently minimal, we also look at the more advanced use of big data in other sectors. But big data alone will not in itself guarantee urban sustainability: supporting policies, including those for reducing energy and private transport use, and improving public health, will also need to be in place.

Chapter 4 sounds a cautionary note about big data applications. In general terms, it discusses, in turn, the potential serious challenges to its use, including privacy, data security, reliability, cost, technical challenges, and potential barriers to its acceptance, which will need to be overcome. The barriers to acceptance and use vary greatly from one application to another, being probably zero for some applications (for example, urban weather forecasting), to possibly serious for more sensitive applications that involve even anonymised personal data. We conclude that big data is not a panacea for all urban problems—some important areas of urban sustainability are probably best tackled by traditional small data approaches or a judicious use of both big and small data. The barriers for some applications, particularly those based on personal data, will for some time be greater in the cities of many industrialising countries than in OECD cities.

Chapter 5 re-examines the general solutions proposed to improve the environmental sustainability of transport discussed in Sect. 1.3.2, with a view to understanding the potential for big data in each of these approaches. How can big data be used to reduce transport energy and emissions in cities? Specifically, how can big data encourage modal shift from cars to more environmentally friendly modes, and reduce vehicular transport overall through better trip planning? The chapter also includes a case study of a 'personal transport planner' designed for use in Beijing, based on the idea of a monthly personal transport energy quota.

Chapter 6 first discusses the *smart grid*, which will be a necessity if electricity production in the future is to be sustainable. The chapter then looks at energy in an urban context, emphasising domestic energy consumption and the role of big data in its reduction. It is found that experience to date shows that data provision alone, for example, that made possible by smart meters, can not on its own effect the large cuts needed in household energy use. However, in future, householders could well be both consumers *and* producers of energy (for example, from rooftop PV cell arrays). Householders will inevitably become far more aware than they are now of the price of electricity and how this varies over time.

Chapter 7 examines the potential for big data in improving urban health and well-being, in the face of the ageing of global society and the rise in real health care costs. It looks at how more use of big data could help solve these and other health challenges, then gives actual or planned examples of its use in healthcare. The Quantified Self movement, discussed next, could prove a forerunner of a more general move to greater patient involvement in monitoring their personal health. The data would come from various apps on their smart phones, wearable devices, or body sensors. The chapter stresses the connection with the transport and energy chapters, given the role of these two sectors in urban air pollution, UHI and global warming and for transport, traffic-related casualties. As a specific example, a case study of a design of an instrumented chair ('Virtual Spine') to improve spinal health and general well-being is included.

Chapter 8 looks to the future, given that applications of big data for urban sustainability are still in their infancy, and it could be many years before it can make a real difference. We try to place big data and urban sustainability problems in the year 2050 or even later, by first describing what the world of 2050 might look like, assuming that nations seriously tackle the global environmental and resource problems the planet increasingly faces. We then explore the possible role of big data in the cities of such a world, both in OECD and non-OECD countries, both in the transition period and later.

References

1. Adomaitis K 2015 Top developed world cities with low reliance on car-based mobility. Accessed on 23 March 2017 at http://blog.euromonitor.com/2015/08/top-developed-world-cities-with-low-reliance-on-car-based-mobility.html
2. Anderson K, Peters G (2016) The trouble with negative emissions. Science 354(6309):182–183
3. Anon (2016) Wildlife loves LA limelight, too. New Sci 230(3070):7
4. Barley S (2010) Escape to the city. New Sci 208(2785):32–35
5. Baum-Snow N, Pavan R (2013) Inequality and city size. Rev Econ Stat 95(5):1535–1548
6. Berger M (2014) The unsustainable city. Sustainability 6:365–374. https://doi.org/10.3390/su6010365
7. Bettencourt LMA, Lobo J, Helbing D et al (2007) Growth, innovation, scaling, and the pace of life in cities. Proc Natl Acad Sci 104:7301–7306
8. Bettencourt LMA (2014) The uses of big data in cities. Big Data 2:12–22. https://doi.org/10.1089/big.2013.0042
9. BP (2017) BP statistical review of world energy 2017. BP, London
10. Carmin J-A, Dodman D, Harvey L et al (2011) Urban adaptation planning and governance: challenges to emerging wisdom. In: Otto-Zimmermann K (ed) Resilient cities: cities and adaptation to climate change. Proceedings of the Global Forum 2010. Springer, New York, NY

11. Coady, D, Parry, I, Sears, S et al 2015. How large are global energy subsidies? IMF Working Paper WP/15/105. Available at https://www.imf.org/external/pubs/ft/wp/2015/wp15105.pdf
12. Cornwall W (2015) Carbon trackers could help bolster climate vows. Science 350(6267):1450–1451
13. Costa N, Matos I (2016) Inferring daily routines from electricity meter data. Energ Buildings 110:294–301
14. Davis LW, Gertler PJ (2016) Contribution of air conditioning adoption to future energy use under global warming. Proc Natl Acad Sci 112:5962–5967
15. Davis SJ, Caldeira K (2010) Consumption-based accounting of CO2 emissions. Proc Natl Acad Sci 107:5687–5692
16. Day JW, Hall C (2016) America's most sustainable cities and regions. Springer Science+Business Media, New York, NY. https://doi.org/10.1007/978-1-4939-3243-6_1
17. Goldewijk KK, Beusen A, Janssen P (2011) Long-term dynamic modeling of global population and built-up area in a spatially explicit way: HYDE 3.1. Holocene 20(4):565–573
18. Gong P, Liang S, Carlton EJ (2012) Urbanisation and health in China. Lancet 379:843–852
19. Energy Information Administration (EIA) (2013) International energy outlook 2013. US Dept. of Energy, Washington, DC. Available at http://www.eia.gov/forecasts/archive/ieo13/pdf/0484(2013).pdf
20. Field CB, Barros VR, Mastrandrea MD et al 2014. IPCC Climate change 2014: impacts, adaptation, and vulnerability: summary for policymakers.
21. Field CB, Mach KJ (2017) Rightsizing carbon dioxide removal. Science 356(6339):706–707
22. Fizaine F, Court V (2015) Renewable electricity producing technologies and metal depletion: a sensitivity analysis using the EROI. Ecol Econ 110:106–118
23. Froggatt A, Schneider M (2015) Nuclear power versus renewable energy: a trend analysis. Proc IEEE 103(4):487–490
24. Grossman L (2017) Nuclear holiday. New Sci 234(3126):20–21
25. Hansen J, Sato M, Hearty P et al (2016) Ice melt, sea level rise and superstorms: evidence from paleoclimate data, climate modeling, and modern observations that 2 °C global warming could be dangerous. Atmos Chemist Phys 16:3761–3812
26. Haubensak O (2011) Smart cities and Internet of Things. In: Michahelles F (ed) Business aspects of the Internet of Things, Seminar of Advanced Topics, FS2011. ETH, Zurich
27. Humphries C (2012) Concrete jungle. Nature 491:514–515
28. Inayatullah S (2011) City futures in transformation: emerging issues and case studies. Futures 43:654–661
29. Intergovernmental Panel on Climate Change (IPCC) (2014) Climate change 2014: mitigation of climate change. Cambridge University Press, Cambridge
30. Intergovernmental Panel on Climate Change (IPCC) (2015) Climate change 2014: synthesis report. Cambridge University Press, Cambridge
31. Intergovernmental Panel on Climate Change (IPCC) (2014) Summary for policymakers. In: Stocker TF, Qin D, Plattner G-K et al (eds) Climate change 2013: the physical science basis. CUP, Cambridge, New York, NY
32. International Atomic Energy Association (IAEA) (2012) Energy, electricity and nuclear power estimates for the period up to 2050. IAEA, Vienna
33. International Council for Local Environmental Initiatives (ICLEI) (2017) Accessed at http://www.sustainable.org/creating-community/inventories-and-indicators/149-international-council-for-local-environmental-initiatives-iclei
34. International Energy Agency (IEA) (2016) Key world energy statistics 2016. IEA/OECD, Paris
35. Keller DP, Feng EY, Oschlies A (2014) Potential climate engineering effectiveness and side effects during a high carbon dioxide-emission scenario. Nat Commun 5:3304. https://doi.org/10.1038/ncomms4304
36. Kleerekoper L, van Esch M, Salcedo TB (2012) How to make a city climate-proof, addressing the urban heat island effect. Resourc Conserv Recycl 64:30–38

37. Kourtit K, Nijkamp P (2013) In praise of megacities in a global world. Region Sci Pol Pract 5(2):167–182
38. Lawson L (2016) Sowing the city. Nature 540:522–524
39. McClellan J, Keith DW, Apt J (2012) Cost analysis of stratospheric Albedo modification delivery systems. Environ Res Lett 7:034019. 8pp
40. McGlade C, Ekins P (2015) The geographical distribution of fossil fuels unused when limiting global warming to 2 °C. Nature 517:187–190
41. Moriarty P (2016) Reducing levels of urban passenger travel. Int J Sustain Transport 10(8):712–719. https://doi.org/10.1080/15568318.2015.1136364
42. Moriarty P, Honnery D (2008) The prospects for global green car mobility. J Clean Prod 16:1717–1726
43. Moriarty P, Honnery D (2009) What energy levels can the Earth sustain? Energy Policy 37:2469–2474
44. Moriarty P, Honnery D (2010) A human needs approach to reducing atmospheric carbon. Energy Policy 38:695–700
45. Moriarty P, Honnery D (2011) Rise and fall of the carbon civilisation. Springer, London
46. Moriarty P, Honnery D (2012) Preparing for a low-energy future. Futures 44:883–892
47. Moriarty P, Honnery D (2012) What is the global potential for renewable energy? Renew Sustain Energy Rev 16:244–252
48. Moriarty P, Honnery D (2015) Future cities in a warming world. Futures 66:45–53
49. Moriarty P, Honnery D (2015) Reliance on technical solutions to environmental problems: caution is needed. Environ Sci Technol 49:5255–5256
50. Moriarty P, Honnery D (2016) Can renewable energy power the future? Energy Policy 93:3–7
51. Moriarty P, Honnery D (2017) Review: assessing the climate mitigation potential of biomass. AIMS Energy J 5(1):20–38
52. Moriarty P, Honnery D (2017) Reducing personal mobility for climate change mitigation. In: Chen W-Y, Suzuki T, Lackner M (eds) Handbook of climate change mitigation and adaptation, 2nd edn. Springer Science+Business Media, New York, NY. https://doi.org/10.1007/978-1-4614-6431-0_73-1
53. Moriarty P, Wang SJ (2014) Low-carbon cities: lifestyle changes are necessary. Energy Procedia 61:2289–2292
54. Moriarty P, Wang SJ (2015) Eco-efficiency indicators for urban transport. J Sustain Dev Energy Water Environ Syst 3(2):183–195
55. Murray J, King D (2012) Oil's tipping point has passed. Science 481:433–435
56. Murray JW, Hansen J (2013) Peak oil and energy independence: myth and reality. Eos 94(28):245–252
57. Nelson B (2013) Natural disasters: a calculated risk. Nature 495:271–273
58. New M, Liverman D, Schroder H et al (2011) Four degrees and beyond: the potential for a global temperature increase of four degrees and its implications. Philos Trans R Soc A 369:6–19
59. Pan W, Ghoshal G, Krumme C et al (2013) Urban characteristics attributable to density-driven tie formation. Nat Commun 4:1961. https://doi.org/10.1038/ncomms2961
60. Pataki DE, Carreiro MM, Cherrier J et al (2011) Coupling biogeochemical cycles in urban environments: ecosystem services, green solutions, and misconceptions. Front Ecol Environ 9(1):27–36
61. Peng S, Piao S, Ciais P et al (2012) Surface Urban Heat Island across 419 global big cities. Environ Sci Technol 46:696–703
62. Pickard WF (2014) Smart grids versus the Achilles' Heel of renewable energy: can the needed storage infrastructure be constructed before the fossil fuel runs out? Proc IEEE 102(7):1094–1105
63. Putz FE, Redford KH (2009) Dangers of carbon-based conservation. Glob Environ Chang 19:400–401
64. Rassia ST, Pardalos PM (eds) (2014) Cities for smart environmental and energy futures, Energy systems. Springer, Berlin. https://doi.org/10.1007/978-3-642-37661-0_2

65. Ravilous K (2010) How climate change could flatten cities. New Sci 208(2782):14
66. Rees WE (2012) Cities as dissipative structures: global change and the vulnerability of urban civilization. In: Weinstein MP, Turner RE (eds) Sustainability science: the emerging paradigm and the urban environment. Springer, New York, NY
67. Rohde RA, Muller RA (2015) Air pollution in China: mapping of concentrations and sources. PLoS One 10(8):e0135749. https://doi.org/10.1371/journal.pone.0135749
68. Ruble BA (2012) The Challenges of the 21st-century city. The Wilson Center, Washington, DC, pp 1–4. Available at www.wilsoncenter.org
69. Satterthwaite D (2000) Will most people lives in cities? Br Med J 321:1143–1145
70. Setälä H, Viippola V, Rantalainen A-L et al (2013) Does urban vegetation mitigate air pollution in northern conditions? Environ Pollut 183:104–112
71. Schindler J (2014) The availability of fossil energy resources' in Factor X: policy, strategies and instruments for a sustainable resource use. In: Angrick M et al (eds) Eco-efficiency in industry and science, vol 29. Dordrecht, Springer, pp 19–38
72. Smith LJ, Torn MS (2013) Ecological limits to terrestrial biological carbon dioxide removal. Clim Change 118:89–103
73. Spreng D (2005) Distribution of energy consumption andthe 2000 W/capita target. Energy Policy 33:1905–1911
74. Statistics Bureau Japan (SBJ) (2015) Japan statistical yearbook 2016. Statistics Bureau, Tokyo. (Also earlier editions). Available at http://www.stat.go.jp/english/data/nenkan/index.htm
75. Steffen W, Richardson K, Rockström J et al (2015) Planetary boundaries: guiding human development on a changing planet. Science 347(6223):1259855. (10 pp)
76. Tanner CJ, Adler FR, Grimm NB (2014) Urban ecology: advancing science and society. Front Ecol Environ 12(10):574–581
77. United Nations (UN) 2014. World urbanization prospects: the 2014 revision. Accessed on 30 November 2016 at https://esa.un.org/unpd/wup/CD-ROM/. Also the 2011 revision.
78. United Nations (UN) 2015. World population prospects: the 2015 revision. Accessed on 15 December 2015 at http://esa.un.org/unpd/wpp/Download/Standard/Population/.
79. Van den Bergh J, Folke C, Polasky S et al (2015) What if solar energy becomes really cheap? A thought experiment on environmental problem shifting. Curr Opin Environ Sustain 14:170–179
80. Weiss M, Patel MK, Junginger M et al (2010) Analyzing price and efficiency dynamics of large appliances with the experience curve approach. Energy Policy 38:770–783
81. Wikipedia 2017. List of countries by vehicles per capita. Accessed on 23 March 2017 at https://en.wikipedia.org/wiki/List_of_countries_by_vehicles_per_capita.
82. Wikipedia 2017. Natural gas vehicle. m.
83. Willis KJ, Petrokofsky G (2017) The natural capital of city trees. Science 356:374–376
84. World Bank 2017. Access to electricity (% of population). Accessed on 21 March 2017 at: http://data.worldbank.org/indicator/EG.ELC.ACCS.ZS?view=chart.
85. Zhou NQ, Zhao S (2013) Urbanization process and induced environmental geological hazards in China. Nat Hazards 67:797–810

Chapter 2
Urban Health and Well-Being Challenges

2.1 Introduction: Global Health and Well-Being Challenges

Nicola Demsey and colleagues [22] have discussed 'the social dimension of urban sustainability', arguing that the term sustainability must include social and even economic dimensions, in addition to the biophysical ones discussed in Chap. 1. Challenges to social sustainability include the perennial (and often inter-related) urban problems of physical and mental health and well-being, unemployment, income and social inequality, crime, and homelessness. This chapter only fully addresses two of these critical urban problems—well-being and health. Nevertheless, it is vital that the solutions proposed for urban physical eco-efficiency and health improvement, at the very least, do not worsen these other problems; ideally, they should support their amelioration.

The remainder of this section examines global health and well-being. The case of OECD cities is explored in Sect. 2.2. Section 2.3, looks at the cities of the industrialising world in general, while Sect. 2.4 examines the important and special case of Chinese cities and the problems that the extremely rapid industrial and urban growth there have caused.

2.1.1 Physical Health and Mortality Worldwide

According to Jonathan Scrutton and colleagues [64], the future of the world's health is being challenged by two developments: first, an ageing society brought about by increases in longevity and falling birth rates, and second, limited resources to pay for rising health expenditures. These authors also argued that globally there is a shortage of healthcare workers. Although on average global longevity is steadily rising, the rise is far from uniform, with some countries even experiencing a decline in longevity [51]. One consequence is that the world's population is also ageing,

© Springer International Publishing AG, part of Springer Nature 2018   23
S. J. Wang and P. Moriarty, *Big Data for Urban Sustainability*,
https://doi.org/10.1007/978-3-319-73610-5_2

with clear implications for health provision, such as the proportion of the population with chronic illnesses. In 1950, with a world population of 2525 million, only 0.56% were aged 80 years or over; by 2015, the corresponding figures were 7350 million and 1.70%. For developed countries, the growth in the aged population was even greater; in Japan, those aged 80 or over comprised 0.44% of the total population in 1950, but 7.80% in 2015, a figure expected to rise to 15.1% by 2050. Already in 2015, there were almost 17 million people worldwide aged 90 years or over [74].

Perhaps not surprisingly, health costs are also rising. The World Health Organization (WHO) [84] data showed that averaged globally, health expenditures (the sum of both private and public costs) were 8.6% of global GDP in 2012, up from 7.7% in 2000. Percentage expenditures were highest for the 'high income' country category (averaging 11.6% of GDP in 2012). The US easily had the highest percentage expenditure in the world—17.0% in 2012, up from 13.1% in 2000. In the US at least, this expenditure as a share of GDP is expected to continue to rise sharply: Font and Sato [27] have projected that by 2050 in the US, health's share will have risen to over 35%, although such a large share is unlikely to be sustainable. Evidently, due to financial constraints alone, the world in future will find it increasingly difficult to maintain present levels of health care, let alone improve them. Radically new approaches to health care are thus urgently needed to contain costs.

At present, most global deaths are the result of non-communicable diseases, for example, the various types of cancer, diabetes, and respiratory diseases. All such diseases today account for 63% of global deaths, and an even higher share in OECD countries [64]. But the health problems facing the populations of Africa, Asia, and Latin America can be very different from those in OECD countries. For a start, infectious diseases are an important cause of death in many countries in these regions, as are vector-borne diseases like malaria and sleeping sickness. In tropical Africa, for example, the United Nations [72] mortality data shows that AIDS, infant mortality from lower respiratory and diarrhoeal diseases, tuberculosis and nutritional deficiency explain much of the 'longevity gap' compared with high-longevity countries.

Low incomes explain much of the differences in the burden of disease and mortality between high- and low-income countries. Income-based differences in health are not only important in low-income countries. A US study [82] compared the mortality data for adults differentiated by education levels, a proxy for socio-economic differences. Their results showed that merely reducing mortality levels of the less educated to that of the better educated group would avoid eight times as many deaths as would the gains from medical advances. Their findings help emphasise the importance of basic public health measures, even in high-income countries. They also show that the strong effects of inequality on health disparities. Another US study found that 'there is as much as a 35-year difference in life expectancy between the healthiest and richest US neighbourhoods and the most ill and deprived' [65].

Other recent US research [31] has found that not only is income inequality growing in the US, but its deleterious effects on health is getting stronger. The study results also suggested that even allowing for income differences, African Americans had poorer health than European Americans. In their case, at least, 'wealth can't always buy health'. But as Case and Deaton [16] have documented, mortality and morbidity are also rising among middle-aged European Americans. Clearly, something is very wrong with the US health system. An earlier study [57] found similar effects for the UK.

Air pollution is a leading cause of global mortality and illness. One study in *Nature* journal estimated that outdoor air pollution alone caused between 1.6 and 4.8 million premature deaths worldwide, mainly in Asia [42]. Estimation of 'premature deaths' with air pollution—whether indoor or outdoor—as a factor is, however, subject to considerable uncertainty [25], as other official estimates have put it as high as seven million [76, 85]. The leading culprit was fine particulate air pollution, especially particles with diameters <2.5 μm ($PM_{2.5}$). The source of these particles varied, but the main source of $PM_{2.5}$ was found to be from domestic combustion for heating and cooking, particularly in China and India. Emission from power stations and vehicular transport were also important in some regions, but in other regions of the world, $PM_{2.5}$ from agriculture was the leading source [42]. Global pollution from $PM_{2.5}$ appears to be still rising: Brauer et al. [13] calculated that it increased 6% in population-weighted terms from 1990 to 2005.

Traffic fatalities are another leading cause of injury and death globally. Global deaths are estimated at around 1.3 million [86], mainly in industrialising nations, with China and India the leading countries. Many millions also suffer injuries of varying severity annually, which can be economically catastrophic for the affected families, because of the combination of loss of income if the victims are wage earners, and medical bills.

Climate change will have major effects on global human health. Existing diseases will usually increase their range, and new diseases will emerge. For example, the spread to higher latitudes and elevations of the *Anopheles* mosquito, the vector for malaria, could well greatly increase the number of annual malaria cases reported—already 200 million in 2014. The new human populations exposed to malaria may have no natural immunity. *Anopheles* and other mosquito species are also the vectors for additional diseases such as yellow fever, dengue, and chikungunya [55]. Moreover, the combination of climate change and increasing urban populations will act together to worsen global health problems, since cities function as incubators for infectious diseases [49]. At an even more basic health level, 'a positive relationship has been observed between regional trends in climate (rising temperatures and declining rainfall) and childhood stunting in Kenya since 1975, indicating that as projected warming and drying continue to occur along with population growth, food yields and nutritional health will be impaired' [46].

2.1.2 Well-Being: Another Component of Health

So far we have only examined illness and mortality. But the World Health Organisation [85] now see health in far broader terms: 'Health is a state of complete physical, mental and social well-being and not merely the absence of disease or infirmity' [34].

There is now a vast literature on 'quality of life' (QOL) indices. Various indices have been developed to measure QOL, and to rank nations and cities globally on this basis [79]. The best known national-level measure is the UNDP's Human Development Index (HDI) which combines three normalised parameters: life expectancy at birth, number of years formal education, and GNI per capita. Each parameter can vary between 0 and 1 in the UNDP's 2010 revision. The HDI of a country (and the index could be readily adapted for cities) is the geometric mean of the three parameters [77]. The HDI can be criticised—only averages are considered, which can hide inequalities across populations in all three parameters, and more fundamentally, why only these three parameters and not others? One notable omission is income differences *within* each country. Average longevity and education level conceal similar disparities within countries. But the HDI (and other suggested indices which aim to improve on the HDI) do provide a readily understood means for roughly comparing welfare in different countries, and for tracking progress in any one country.

Income distribution inequalities are widening even as average per capita incomes, measured on a national basis, are converging. In a number of major OECD economies, including the US, Japan and many of the countries of Western Europe, this increasing inequity has taken the form of a steady rise in the share of both wealth and income going to the top one percent of families or income earners. Declining top tax rates has been an important reason for their gains [5]. It suggests that this elite is gaining increasing control over the political process in these countries. Income inequality is also rising in most non-OECD countries; in China, there is also rising regional inequality. The effect of this inequality for cities in various regions will be considered further in each of the following sections.

A specifically *urban* index of QOL is the Mercer Quality of Living Survey, which ranks 221 of the world's cities. Singapore, ranked at 26th, is the only city outside the OECD to be placed in the top 50 cities [47, 78]. The aim of the Mercer survey is to help companies decide on remuneration for their employees working overseas, and so is only indirectly concerned with how ordinary residents view their urban QOL. Mercer does now offer an *infrastructure* rankings, which are based on transportation infrastructure, water quality and reliability of electricity service. Again OECD cities dominate the highest rankings, with the important exception of Singapore, which was ranked first in 2017. The reasons for the relative absence of non-OECD cities, even some high-income Asian cities—in Mercer's top 50, will become clear from the discussion in Sects. 2.3 and 2.4. Looking just at US cities, one economically-oriented study found that neither large urban populations nor high densities decreased the QOL for residents [3].

2.2 Urban Health in OECD Cities

In OECD countries, the health of urban residents overall is usually better than those living in rural areas, mainly because of higher levels of medical services and higher incomes. (Nevertheless, a study in the UK [61] found that urban residents suffer more from mental health problems than do rural residents). For the US, Susan Blumenthal and Jessica Kagen [11] have cautioned that low-income residents have poorer than average health, regardless of where they live. One review study reported that, if the Gini coefficient (a measure of income inequality) was reduced below 0.3 for 30 OECD countries, 1.5 million annual deaths could be avoided [58]. This finding suggests that lessening social and economic inequalities is a important approach for improving overall health levels.

OECD cities will face several major health challenges in the future. Three, already discussed, concern all areas: an ageing population, rising health costs, and new diseases (and the geographical spread of existing ones) because of climate and land use change. But health threats that are far more pronounced in urban than rural areas include the still serious health effects of air pollution—and also noise pollution. In future, ongoing climate and global land use change could cause worsening air pollution levels, even at constant annual pollutant emissions to the atmosphere.

2.2.1 Urban Air and Noise Pollution

It might be thought that air pollution is no longer a problem in the cities of the mature industrialised countries, that the soot and lead pollution problems that are too often a feature of cities in the industrialising world are largely a thing of the past. It is true that emissions of sulphur oxides have been dramatically reduced, following deadly air pollution (smog) episodes in Donora, a mill town in Pennsylvania USA in 1948, where hundreds died, and in London in 1952, which killed an estimated 4000, and made 100,000 ill. (In London in 1952, $PM_{2.5}$ levels rose to 3000 μg m^{-3} [44], far in excess of even the worst levels in today's Beijing or Delhi.) Both episodes were an important impetus for air pollution legislation in those countries. But two remaining air pollutants are proving far harder to reduce: very small particulates and oxides of nitrogen.

Although diesel fuel has long been the main fuel for heavier road vehicles, its use for passenger vehicles has lagged, particularly in the US. In Europe, however, diesel-fuelled cars have been promoted because of their better fuel efficiency than petrol vehicles. But the health problems arising from particulate emissions, particularly particles <2.5 μm ($PM_{2.5}$) emitted by diesel vehicles has led the city government in Paris—and France leads Europe in the diesel share of the car fleet—to enact a ban on diesel-fuelled cars beginning in 2020 [56]. Further, with the emphasis on RE and especially bioenergy in the EU, wood-burning stoves are becoming popular. But even correctly installed ones can deliver very high levels of $PM_{2.5}$ for the house occupants—even much greater than living on a heavily polluted street [41]. These microscopic particulates have greater health effects than larger ones—they cause greater levels of 'oxidative stress' in cells. It may even be the case that exposure to $PM_{2.5}$ is implicated in increased risk of dementia and Alzheimer's disease [71].

In London, the legal EU pollution limits for NO_2 (an air pollutant causing heart and lung problems) for the whole of year 2017 were already exceeded by 5 January 2017 [6]. Overall in the UK, [40] particulate emissions are estimated to be responsible for 30,000 deaths annually, and NO_2 for 10,000.

Noise pollution is increasingly recognised as a health hazard in its own right. Like air pollution, it is not a new phenomenon—ancient Rome reportedly banned carts from its streets at night because of their noise. A 2011 WHO study on noise effects in Europe [83], found that noise exposure from road traffic increased the risk of heart disease. Both road and aircraft noise increased the risks from high blood pressure. There was also some evidence that school children subject to prolonged noise suffered cognitive impairment, which persisted for 'for some time' after the noise exposure stopped. Other effects found included annoyance and sleep disturbance.

The high population densities found in many cities, particularly in the central and adjacent areas, exacerbate the impact of air and noise pollution. Not only do the high density of residents and workplaces generate high levels of passenger and freight traffic, but the resulting traffic congestion increases both noise and air pollution emissions per km of vehicular movement, and the large numbers of exposed people living, working or moving through such areas increases the absolute health impact of these pollutants.

Urban health is not independent of the other sectors we have discussed in Chap. 1. Apart from the obvious health implications of traffic casualties, transport adversely affects the health of urban residents, particularly those living near heavily trafficked highways through air pollution and noise emissions. A further impact arises from the choice of travel mode. Not only can shifting from motorised to non-motorised travel, through the exercise it provides, improves the health of the individual non-motorised traveller, but by helping to reduce air and noise pollution, all non-motorised travellers confer health benefits on other urban residents. In fact, there is considerable synergy between efforts to reduce air pollution and climate mitigation and adaptation efforts [32].

2.2.2 Climate Change Effects on Urban Health

The combined effects [43] of UHI (see Sect. 1.3) and climate change-related increases in duration and frequency of heatwaves in some regions have already resulted in thousands of heat-stress deaths, particularly in Europe; the 2003 European and 2010 Russian heatwaves are estimated to have caused excess mortality of 80,000 and 54,000 respectively [81]. Mortality occurred disproportionately in cities, and among the elderly [52]. Since the global population is ageing, any future increase in either the frequency or severity of heatwaves, urban or not, is of concern. Camilo Mora and colleagues [50] reviewed numerous studies on heat waves globally and found that, at present, about 30% of the global population experienced climatic conditions that lead to excess mortality from heat waves for at least 20 days each year. By 2100, this percentage could rise to 48% (strong GHG reductions) to as high as 74% (GHG emissions continue their growth).

If the world follows the business-as-usual scenarios (RCP6.0 and RCP8.5) of the IPCC, global temperature rises of 3 °C or more above pre-industrial can be expected by the end of this century [35]. Prolonged outdoor activity would then be dangerous, and, in the absence of air-conditioning, even indoor living. If fossil fuels were continued to be used for electricity, the resulting greatly increased use of air-conditioning would worsen climate change at the global level. We thus need to be careful that local urban climate mitigation or adaptation solutions are not at the expense of solutions at the global level.

Along with many other areas, OECD cities will see the spread of diseases like malaria to formerly temperate areas as temperatures rise (see Sect. 2.1.1). Temperature increases, whether caused by UHI effects or climate change, are also forecast to increase levels of some pollutants in urban areas. Levels of air pollution depend not only on the emissions per day or per year, but also on atmospheric conditions. Favourable conditions can rapidly disperse urban airborne emissions, but conditions such as temperature inversions can prevent this, thus leading to rising urban concentrations, and with it adverse health effects from such pollution.

Dominique Charron and colleagues [17] have documented the incidence of waterborne diseases in North America, noting how extreme rainfall events can increase the risk of enteric diseases. Since increases in extreme rainfall, floods, and higher temperatures can all be expected under future climate change, enteric disease risk can also be expected to rise. The risk of food-borne diseases is also likely to rise following climate change. For instance, *Campylobacter* is the most frequently reported gastrointestinal disease in Europe, with contaminated poultry the main source of human infection (40–70% of cases). Kovats et al. [38] reported that increasing seasonal temperatures lead to a corresponding rise in the number of humans infected with *Campylobacter*.

2.2.3   Stress and Mental Illness

It is not always easy to place physical and mental illness in separate compartments. Loneliness, for example, is thought by some researchers to be implicated in several physical diseases [48]. Although temporary feelings of loneliness are common, chronic loneliness may affect the 'cardiovascular immune, and nervous systems'. This, in turn, could help explain why the longevity of socially isolated people are found to be lower than others. Nor is mental illness an isolated problem: in the US, a 2017 report found that levels of mental illness were at their highest level for at least 60 years [7]. According to the survey, some eight million US adults were experiencing mental health problems.

It may even be the case that city life itself is making us ill, that the term 'big city disease' is more than a metaphor. City living has been found to be associated with a variety of mental illnesses, including schizophrenia [39]. A long-term longitudinal study of mental health in Camberwell, a London suburb, found that over the period 1965 to 1997, the incidence of schizophrenia per 100,000 population had roughly doubled. However, no such rise was found in the general UK population [1]. Since most of the global population are now already urban, and the UN [73] expect this share to rise, any link between city living and mental illness is worrying. The link is unproven, even if it is easily explained: 'City dwellers typically face more noise, more crime, more slums and more people jostling on the streets than do those outside urban areas. Those who have jobs complain of growing demands on them in the workplace, where they are expected to do much more in less time' [1]. One theory is that the stress of city living increases the risk of mental illness mainly in people who are already susceptible because of other life problems or because of their genetic makeup.

One intriguing idea is that not only urban stress, but even crime and levels of aggression, can be ameliorated by the provision of more green spaces in the city [28]. In the US, researchers [8] surveyed a range of urban and rural environments in the state of Wisconsin. They found that, after adjustment for length of residence: 'Higher levels of neighborhood green space were associated with significantly lower levels of symptomology for depression, anxiety and stress, after controlling for a wide range of confounding factors'. *Local* provision of green space may be especially important—and effective in improving mental well-being—for those with limited ability to travel outside their immediate area, such as the elderly and low-income groups. Julie Dean and colleagues [20] go further and have hypothesised that restoring biodiversity in urban areas is important for mental health.

Extreme weather events can also adversely affect mental health [26], suggesting that, *ceteris paribus*, cities can expect levels of mental illnesses to rise as climate continues to change.

2.2.4 Discussion

So far in this section, we have looked in turn at urban pollution as a present serious urban health hazard, and ongoing climate change, which will have increasingly important future influence. We then examined whether urban life itself and the stress it generates is at least partly responsible for the rising mental illness levels recorded in many cities, and discussed the role of urban vegetation and green spaces as a remedial measure.

A very different explanation for much illness, whether physical or mental, is income and social inequality. For physical health, much of the health disparity is no doubt mediated by different pollution levels, access to health services, and lifestyle factors experienced by different income classes. But for mental health, much evidence suggests that inequality itself has a largely unacknowledged deleterious effect [58].

2.3 Urban Health and Liveability in Non-OECD Cities

The health of those living in urban areas is recognised as presenting its own specific problems, with a dedicated journal: *The Journal of Urban Health* [21, 75]. For the very poor in cities in tropical Africa and parts of Asia, the key urban health problems may stem from malnutrition, lack of clean drinking water, and poor sanitation. These basic problems, which can account for a high proportion of deaths in low-income cities [86], will obviously not be solved by providing more hospitals, medicines, or health care professionals. Nevertheless, other health problems are also important, including established contagious diseases such as AIDS, and the threat of emerging diseases such as Ebola in West Africa, and Zika in Latin America.

Many low- and middle-income countries are undergoing a transition in mortality causes. For the West African city of Accra, Ghana, Agyei-Mensah and de-Graft Aikins [2] have discussed what they termed the 'epidemiological transition' that the city has experienced as it modernised over the past century. They found that the protracted shift from infectious diseases to chronic diseases varied according to class, with wealthier residents increasingly likely to experience morbidity and mortality from chronic diseases such as hypertension and diabetes. For poorer residents, infectious diseases, particularly AIDS in recent decades, were more important, but chronic illnesses were also now an additional health burden. They identified three factors as important for their 'protracted polarised model of epidemiological transition in Accra: urbanization, urban poverty, and globalization.'

Cities everywhere run higher risks from epidemics than do non-urban areas. There are several reasons for this. In the industrialising countries of Africa and Asia, migration to cities is proceeding apace, and in all countries, travel between cities is growing; both factors help make cities centres for infection [4]. Also, cities are by definition densely populated, which facilitates the spread of infectious diseases, including HIV [52].

In 2006, Mike Davis published an article called '*Planet of Slums*' [19]. In many of the cities of the industrialising world, informal settlements—slums, *favelas*, *bidonvilles*—are the home of a significant share of the urban population, both new urban migrants, and locally born residents. The UN has estimated that 863 million people, or about one-third of the urban residents of the industrialising world, live in such slums [24]. Figure 2.1 shows two such slums, one from Kenya and one from Brazil, with the latter illustrating the steep slope locations characteristic of many Latin American slums. The number of slum dwellers is likely to swell as migration to cities continue, whether because of rural poverty, drought or political instability. Ebola was formerly limited to rural regions, but in the 2014 outbreak, the cities of the relevant West African nations were not spared. Moreover, in these cities, the urban poor were the worst affected. Slums also often lack clean drinking water and basic sanitation, creating ideal conditions for the spread of other infectious diseases such as typhoid.

The rest of this section examines the effects of both air pollution and climate change on the health of urban residents in cities outside the OECD. The case of Chinese urbanisation is so important it is considered separately in Sect. 2.4.

2.3.1   Urban Air Pollution and Climate Change Effects

Air pollution has long been a problem for urban settlements everywhere. Millennia ago, the main causes were indoor fires and in some cases, industries such as lead smelting. But a combination of growing levels of vehicular transport, vastly increased industrial activity, and lax enforcement of anti-pollution legislation (even assuming it exists) has seen air pollution rise to unprecedented levels in many cities

Fig. 2.1 Slums. (**a**) Kibera slum in Nairobi, Kenya. (**b**) *Favela*, Rio de Janiero, Brazil ((**a**) ['Scenes from Kibera slum in Nairobi' by khym54, available at http://bit.ly/2uWr3o2 under a Creative Commons licence 2.0 Generic. Full terms at http://creativecommons.org/licenses/by/2.0.]. (**b**) 'Favela, Rio de Janiero' by Eloise Acuna, available at http://bit.ly/2eGEVeR under a Creative Commons licence 2.0 Generic. Full terms at http://creativecommons.org/licenses/by/2.0.])

of the industrialising world. For example, Delhi, the world's most polluted megacity, has average annual levels of $PM_{2.5}$—the particle size which causes the most serious health problems—of 122 micrograms per cubic metre (122 µg m^{-3}) two orders of magnitude above the 10 µg m^{-3} WHO standard [68]. In winter, pollution levels can be even worse, rising as high as 600 µg m^{-3}, the result of a combination of domestic fires for heating and cold-weather temperature inversions.

In many industrialising countries, exposure to high concentrations of *lead* is an important health problem, particularly in urban areas. The risk is greatest for children, as it can affect their neurocognitive development. Although leaded fuels have been phased out in OECD countries, such is not the case in many other countries, especially in Africa. Other sources of lead in the environment include various industries such as smelting and battery manufacturing. The lead from such fuels and industries finishes up in the atmosphere, dust, and soils. Lead pollution is of serious concern in the industrial areas and cities of other countries as well, such as China, Bangladesh, India, and Mexico. For example in Dhaka, the megacity capital of Bangladesh, 'lead concentrations in airborne particulate matter averaged 453 ng m^{-3} during the low rainfall season of November to January' [69].

Ongoing climate change is anticipated to have significant effects on health everywhere, with the effects rising with temperature. Many of these impacts, such as water- and food-borne diseases, are already major problems in poor cities. Another crucial impact could be rising malnutrition, the result of extreme weather events like drought, but also of declining yields for key crops expected in hot regions as plants move closer to their physiological tolerance limits [35]. The various health impacts, like other impacts, however, will be highly uneven; poor households with correspondingly low carbon footprints have contributed least to global warming but will experience the greatest deleterious effects [15].

Further, the modelled results of Jeremy Pal and Elfatih Eltahir [54] suggest that by the end of this century, the human adaptation limits to temperature may be reached in cities in the Persian Gulf region, even under business-as-usual scenarios for climate change. At wet bulb temperatures (which combine temperature and humidity levels) above 35 °C, humans can no longer adapt to the heat by perspiration evaporation. At even higher temperature increases, large regions of the world would become uninhabitable [66]. Matthews et al. [45] have argued that even if we are successful in limiting global temperature rises to 2 °C, a number of the worlds megacities will experience deadly heat stress, similar to that occurring in 2015 in Kolkata and Karachi. Among the newly affected megacities would be Lagos and Shanghai, even if the temperature rise above pre-industrial was limited to 1.5 °C, as advocated at the Paris climate conference.

2.3.2 Mental Health, Well-being and Liveability

For the very poor in cities, the struggles to meet such basic needs as housing, adequate nutrition, and medical care are not the only problems they face. Many also suffer from a range of mental and social illnesses. Writing from a South Asian context, Trivedi and colleagues [70] have provided the following list: 'psychoses, depression, sociopathy, substance abuse, alcoholism, crime, delinquency, vandalism, family disintegration, and alienation.' The authors stressed that poverty and mental illness cannot be easily separated. Further, lower-income residents are not only more likely to turn to crime, but also to be its victims. For the case of the megacity Delhi, Gautam Bhan [9] has even referred to the 'impoverishment of poverty' when discussing the evictions of the poor from illegal settlements, and their *de facto* 'criminalization' and loss of any rights as urban citizens.

2.4   Urban Health and Liveability in Chinese Cities

2.4.1   Rapid Urbanisation in China

There are several features of urbanisation in China which make it unique and are important for the health and well-being of urban residents there. First is the sheer scale of urban resident numbers. According to the UN [73, 74], an estimated 787 million resided in cities, 20% of the global urban population. China in 2015 had some 115 cities greater than one million population. The second remarkable feature is the *rate* of urbanisation in recent decades. The historic urban growth rate has been very uneven, as Fig. 2.2 shows. Although urbanisation began in China four millennia ago [14], the country was only 11.8% urban by 1950. It rose during the Great Leap Forward beginning in the late 1950s, but the urban share actually fell during the ensuing Cultural Revolution of 1965–1975. Since the opening up of the economy, the rate has sharply accelerated, and the urban share has more than doubled since 1990 to 55.6% in 2015, slightly above the global average of 54% [73]. From another point of view, the urban built-up land area in China expanded by 513% between 1981 and 2012, from 7438 to 45,566 km^2, with an annual growth rate of 6% [18].

With average annual national urban growth rates of around 3%, and national population growth rates of only 0.6%, it is clear that the growth of China's cities is fuelled largely by massive migration from rural areas. So urban growth and change has been paralleled by corresponding population change in rural areas, and the accompanying social changes. Moreover, this trend is expected to continue: from 2010 to 2025, according to the Ministry of Housing and Urban-Rural Development in China, the plan is to transfer a further 300 million people who are now living in rural areas to urban areas. In this migration from rural areas to cities, China is similar to many other countries in Asia and Africa [73], and it produces the same problems for the newcomers to cities. The recent rate of growth of urbanisation in China is unprecedented; in Western Europe, for example, it took the entire eighteenth century for the urban population to grow from 21.4% to 40.6% [29]. Put another way, for the past decade or more, China has been adding the equivalent of a new Chicago to its urban population every *month* [62].

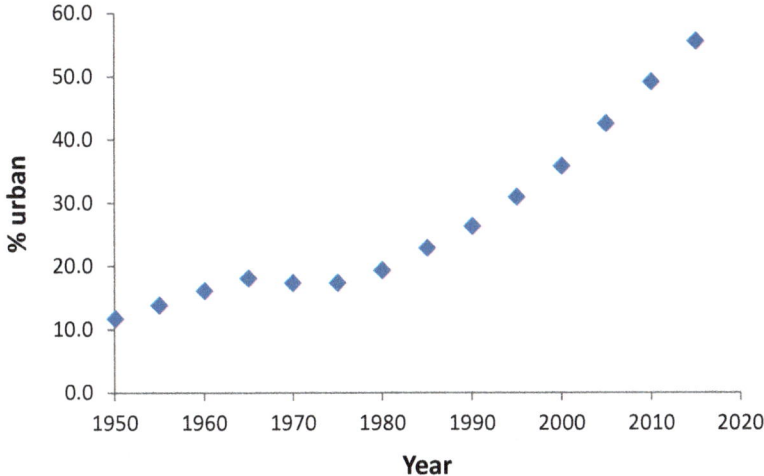

Fig. 2.2 Urbanisation in China, 1950 to 2015. Source: [72]

The third distinct feature of Chinese urbanisation is the uncertain status of the rural migrants to cities. Already in 2008, these numbered 140 million, or 10% of the national population, a number expected to grow [33]. These migrant workers, who tend to be young, male, and single, cannot become full urban citizens; they live in a semi-urban state, they do not enjoy (or only partially enjoy) the rights urban residents have for national education, health, social security and assistance, and housing security. Nor do they enjoy the right to vote or to be elected to public office. If their number was deducted from China's urban population, China's calculated urbanisation rate would be much lower.

Not only are these migrants ineligible for public health insurance, but their health profiles are different from the rest of the urban population. They have a greater incidence of infectious diseases such as 'acute respiratory infections, diarrhoeal, parasitic, and sexually transmitted diseases, and tuberculosis' [33]. The rates of innoculation for diseases such as tuberculosis and measles for the children of such migrants is also lower than either that of their fellow urbanites—or even that for rural regions [30].

Given the risk of spread of these diseases, the general health of cities requires that their plight be given priority. Nevertheless, the migrant workers tend to be younger and healthier than the general rural population when they arrive in cities. But if they get seriously ill through disease or industrial accidents, they are likely to return home for family support to avoid the high medical costs they would incur. As Hu et al. [33] have summarised: 'In essence, the countryside is exporting good health and reimporting ill-health.' In countries like China that are rapidly urbanising, the health of rural and urban populations are interrelated.

Yet the notion of 'urbanisation' itself can be problematic, and nowhere is this more apparent than in contemporary China. Christian Sorace and William Hurst [67], in an article that examined the creation of 'ghost cities' in China, have contested the idea that rural to urban migration is the main driver of urbanisation. They believed that, unlike the case in tropical Africa, the Chinese government actively encourages the rapid urbanisation of China.

2.4.2 Air Pollution in Chinese Cities

The massive industrialisation and urbanisation of China in recent decades has been accompanied by serious pollution of China's air, waters and soils. The Chinese leadership has repeated stressed the need to push forward with urbanisation. If the vast majority of the Chinese people are to live in urban centres, it is thus of great importance to find ways to develop healthy cities, to avoid and overcome the 'big city disease'.

Urban residents in China regard air pollution as a key problem, and the findings back them up. In China overall, total deaths from air pollution were recently calculated at 1.6 million annually [62]. For cities, the main cause of anthropogenic air pollution is rising energy production, which is concentrated in urban and peri-urban areas. China is by far the world's largest coal producer, consuming exactly half the world's total in 2016 [12]. Not surprisingly, coal accounts for nearly two-thirds of energy consumption, although its use fell slightly from 2014 to 2015. Consequently, China is not only the world's largest emitter of CO_2, as well as of SO_2, but is also a major emitter of other GHGs, methane, and carbon black [12, 37]. Much of this coal is still used for domestic heating in urban and other areas. The resulting air pollution mainly consists of coal smoke, especially fine particulate matter and SO_2.

Beijing, China's capital, is one of the world's most polluted cities (see Fig. 2.3). A record was set on 12 January 2013, when $PM_{2.5}$ levels rose to 886 μg m^{-3} in Beijing in the evening. That winter day's average was 22 times the WHO safe level, and over January, authorities advised residents to stay indoors. SO_2 emissions have fallen since 1999, because of SO_2 scrubbers on power stations. But, in addition to $PM_{2.5}$, NO_2 levels remain high. The region's unfavourable topography can hem the smog in over winter months, when domestic coal burning is at its highest [44].

Starting in 2012, however, China has progressively introduced a nation-wide air quality reporting system, and by 2015 this had already covered 945 sites in 190 cities. 'These automated stations report hourly via the Internet, and focus on six pollutants: particulate matter <2.5 μm ($PM_{2.5}$), particulate matter <10 μm (PM_{10}), sulfur dioxide (SO_2), nitrogen dioxide (NO_2), ozone (O3), and carbon monoxide (CO)' [62]. These cities already make this air quality data immediately available to the public, and the plan is that the rest of China's 338 major cities will soon follow. Beginning in 2014, the public will also be able to access on the Internet the real-time air emissions and discharges into waterways from over 15,000 national heavy polluters [60].

Accurate assessment of pollution levels in cities is urgently needed. Also needed, of course, are strong anti-pollution policies. In 2014, the Chinese government announced two important changes to combat urban and regional air pollution. First, a shift from coal as the dominant energy source to cleaner burning natural gas and renewable energy. Second, the implementation of tougher controls on emissions.

Fig. 2.3 Pollution in Beijing, China ['Beijing Air Pollution … ' by Kentaro Iemoto available at http://bit.ly/2tSsaQo under a Creative Commons licence Attribution-ShareAlike 2.0 Generic. Full terms at http://creativecommons.org/licenses/by/2.0.]

Such implementation is anticipated to meet resistance from industry, as it will add to their production costs and partly erode the cost advantage of China's manufactured exports. Three city regions have been targeted: Greater Beijing, and the Yangtze and Pearl River Deltas, each with specified emission reductions [60]. Regions, rather than individual cities are used because of shared air basins.

Even with the high pollution levels in Chinese cities, the health of urban residents is better than their rural counterparts, because of both higher incomes and better medical services—and because $PM_{2.5}$ levels can also be high in rural areas [13, 87].

2.4.3   Liveability in Chinese Cities

There are other encouraging signs that the Chinese government is recognising the serious problems presented by China's hyper-growth in urbanisation and industrialisation. The National New Urbanization Plan (2014–2020) released by the Central Government proposed a shift in thinking about cities 'from land-centered urbanization to people-oriented urbanization.' For such an approach to succeed, the general public needs to be actively involved, but must also realise that the area available for urbanisation in China is limited [18].

Unlike the usual environmental sustainability concept, liveability pays special attention to the living conditions of individuals in a compact urban environment, especially urban apartment residents in cities dominated by speculative developers. Liveability can suffer if residents must live at high population density, surrounded by other tall buildings, and with little access to green areas. Thus, while the general public may support the principle of sustainable urban development, many argue that high-density development is too costly for the individual quality of life. Ensuring sustainability is often considered a reason for urban intensification, but a very narrow definition of sustainability, with its focus on the physical environment and energy aspects, is not that commonly used in the context of personal life.

Yet the goal of liveability is concerned with how we develop and build the future urban environment, which has the function of guiding our actions, planning and re-designing functions that will support people who come from different regions, with different backgrounds, different cultures and living habits to lead a full life in the same city environment. The emphasis on the liveability is becoming more and more important at all levels to direct the future urbanisation development process.

2.4.4 Zhuhai City Experiences

Here we describe the personal experience of Mr. Zhang (not his real name), a rural migrant who has witnessed the vast changes in one city, Zhuhai, over the past two decades. Zhuhai, a city located in the Pearl River Delta has Macau on its southern border. The northern area of Zhuhai is now part of the heavily populated Guangzhou-Shenzhen built-up area, with about 45 million inhabitants. Zhuhai, with its many islands, is an important tourist destination.

When Zhang arrived in Zhuhai for the first time in 1997, it was a typical Chinese town that had just started its urban development. He still remembers that Zhuhai Road was 'very wide' at that time, and that there were many places where you could 'still smell the taste of the countryside, which means the synthesis smell of green plants, grass, manure, and haystacks, mixed with the taste of cooking smoke'. Since then, he has witnessed the city's rise at an increadible speed—'now, Zhuhai's private cars have become too many, and it seems to me the road has become much narrower' (Fig. 2.4).

'It should be there, but now I can't find it', he continued, standing in the middle of a cluster of skyscraper buildings. Zhang pointed that 'before, our school was in the back of that building; now it has been removed, moved to a more distant place from the city'. Despite Zhang's misgivings, Zhuhai was ranked as China's most liveable city in a 2014 report by the Chinese academy of Social Sciences [80].

'Suddenly, the crane seemed to emerge from the ground, and there were cranes everywhere,' Zhang continued. 'The whole of the Golden Bay is a construction site, and many of the skyscrapers quickly filled the entire development zone of the Pearl River and are full of new small office and home office buildings.'

Fig. 2.4 Zhuhai city, China ['Foggy day, Zhuhai city skyline China' by Chris, available at http://
bit.ly/2uqa2iN under a Creative Commons licence Attribution – ShareAlike 2.0 Generic. Full
terms at http://creativecommons.org/licenses/by/2.0.]

'You will see a lot of young workers coming here, especially in the construction
industry, because the wages are relatively high, but they say that they don't enjoy
much of the life here, they will return to their hometown once they can make enough
money. For these workers, there is nothing much else to do except work …only
construction sites, factories and dormitories. The building site, their dormitory, and
a wall are everything in their life … these migrant workers are really poor'. He
shook his head sadly and recalled that: 'A few years ago, I had a relative from the
country, looking for opportunities in Zhuhai, but eventually he became desperate
and felt that there was no room for him.'

2.5 Discussion

The present century will not be an easy period for improving world health and well-
being, because of several adverse factors. First global population is still growing,
and could rise to over 9.7 billion by mid-century [74]. Two-thirds of this number
are anticipated by the UN to live in cities [53], and Ruble [63] even claims the
urban age is just beginning. If present trends continue, many of these will live in
slums under very crowded and unsanitary conditions. By 2050, about 60% of
humanity will live in the tropics, where the disease burden is higher than in temperate climates. The global population is ageing; older people are more susceptible to
infectious diseases. Finally, air travel is expected to grow rapidly (see Sect. 8.1),
with most of the growth expected in non-OECD countries [10].

All these factors favour the spread of newly emerging (or re-emerging) infectious
diseases, mainly viral. Control of bacterial diseases (e.g. the Haiti cholera outbreak
in 2010) has been made more difficult by the emergence of anti-biotic resistance,
largely through the over-use of antibiotics [10]. Recent years have already witnessed
outbreaks of new diseases, including the on-going Zika disease in South America,

Ebola in West Africa in 2014, severe acute respiratory syndrome (SARS), avian flu (H5N1), and swine flu (H1N1).

Climate change will also exacerbate existing urban health problems. Rising temperatures, coupled with the UHI effect, will increase the frequency and severity of heat waves. In many cities, climate change will worsen air pollution, and so its impact on urban health. Existing diseases could see their range spread to new areas. In many poor countries, climate change could adversely affect agricultural production, leading to forced migration to cities that already cater poorly to the health needs of existing residents. As global temperatures rise, a rising share of the global population will be subject to increased simultaneous risks from a variety of impacts arising from changes to water supply, ecosystems, agriculture, and disease, among others [59].

We have seen that the cities of low-income countries have more intractable health problems than OECD cities. Yet a WHO researcher, Christopher Dye [23] has provocatively argued that the main barriers to better urban health in low-income cities are not technical, nor even financial. Instead, the barriers arise from poor governance and policies for public health improvement, and a lack of social inclusion. With better access to modern medicine, higher income households can to some extent insulate themselves from the diseases of lower income households, and so ignore their health problems. The example of Cuba is instructive: longevity there in 2013 was 78 years, similar to the US value at 79 years. In 2014, GDP per capita in Cuba was only $6368, compared with $50,621 in the US ($US 2010 values) [36, 84], which adds support to Dye's argument.

This chapter has outlined the existing health and well-being problems that present urban dwellers face, and how climate and other changes will likely increase the difficulty in providing adequate health standards for all. Clearly, new ideas are needed. In Chap. 7, we will examine possible solutions to these challenges, and the many ways that application of big data can help, provided the political will is there for public health improvement, and basic human needs are first met.

References

1. Abbott A (2012) Urban decay. Nature 490:162–164
2. Agyei-Mensah S, de-Graft Aikins A (2010) Epidemiological transition and the double burden of disease in Accra, Ghana. J Urban Health 87(5):879–897
3. Albouy D 2008. Are big cities bad places to live? Estimating quality of life across metropolitan areas. NBER Working Paper 14472. http://www.nber.org/papers/w14472.
4. Alirol E, Getaz L, Stoll B, Chappuis F, Loutan L (2010) Urbanisation and infectious diseases in a globalised world. Lancet Infect Dis 10:131–141
5. Alvaredo F, Chancel L, Piketty T et al 2017. Global inequality dynamics: new findings from WID.World. NBER Working Paper 23119. Available at http://www.nber.org/papers/w23119.
6. Anon (2017) London breaks pollution limit. New Sci 233(3108):6
7. Anon (2017) Mental distress. New Sci 234(3122):6
8. Beyer KMM, Kaltenbach A, Szabo A (2014) Exposure to neighborhood green space and mental health: evidence from the survey of the health of Wisconsin. Int J Environ Res Public Health 11:3453–3472

9. Bhan G (2014) The impoverishment of poverty: reflections on urban citizenship and inequality in contemporary Delhi. Environ Urban 26(2):547–560
10. Bloom DE, Black S, Rappuoli R (2017) Emerging infectious diseases: a proactive approach. Proc Natl Acad Sci 114:4055–4059
11. Blumenthal SJ, Kagen J (2002) The effects of socioeconomic status on health in rural and urban America. JAMA 287(1):109
12. BP (2017) BP statistical review of world energy 2017. BP, London
13. Brauer M, Amann M, Burnett RT et al (2012) Exposure assessment for estimation of the global burden of disease attributable to outdoor air pollution. Environ Sci Technol 46:652–660
14. Calabro J (2012) Chinese urbanization: efforts to manage the rapid growth of cities. Global Major E J 3(2):75–85
15. Campbell-Lendrum D, Corvalán C (2007) Climate change and developing-country cities: implications for environmental health and equity. J Urban Health 84(1):i109–i117
16. Case A, Deaton A (2015) Rising morbidity and mortality in midlife among white non-Hispanic Americans in the 21st century. Proc Natl Acad Sci 112(49):15078–15083
17. Charron D, Thomas M, Waltner-Toews D et al (2004) Vulnerability of waterborne diseases to climate change in Canada: a review. J Toxicol Environ Health 67:1667–1677
18. Chen M, Liu W, Lu D (2016) Challenges and the way forward in China's new-type urbanization. Land Use Policy 55:334–339
19. Davis M (2006) Planet of slums. New Perspect Q 23(2):6–11
20. Dean J, van Dooren K, Weinstein P (2011) Does biodiversity improve mental health in urban settings? Med Hypotheses 76:877–880
21. de Leeuw E (2012) Do healthy cities work? A logic of method for assessing impact and outcome of healthy cities. J Urban Health 89(2):217–231
22. Dempsey N, Bramley G, Power S, Brown C (2011) The social dimension of sustainable development: defining urban social sustainability. Sustain Dev 19:289–300
23. Dye C (2008) Health and urban living. Science 319:766–769
24. Eisenstein M (2016) Poverty and pathogens. Nature 531:S61–S63
25. Fleming N (2017) Cutting through the smog. New Sci 234(3124):35–39
26. Fleming E, Haines A, Golding B et al (2014) Data mashups: potential contribution to decision support on climate change and health. Int J Environ Res Public Health 11:1725–1746. https://doi.org/10.3390/ijerph110201725
27. Font JC, Sato A (2012) Health systems futures: the challenges of technology, prevention and insurance. Futures 44:696–703
28. Gilbert N (2016) A natural high. Nature 531:S56–S57
29. Goldewijk KK, Beusen A, Janssen P (2011) Long-term dynamic modeling of global population and built-up area in a spatially explicit way: HYDE 3.1. Holocene 20(4):565–573
30. Gong P, Liang S, Carlton EJ et al (2012) Urbanisation and health in China. Lancet 379:843–852
31. Hamzelou J (2017) Wealth can't always buy health. New Sci 234(3121):11
32. Harlan SL, Ruddell DM (2011) Climate change and health in cities: impacts of heat and air pollution and potential co-benefits from mitigation and adaptation. Curr Opin Environ Sustain 3:126–134
33. Hu X, Cook S, Salazar MA (2008) Internal migration and health in China. Lancet 372:1717–1719
34. Huber M, Knottnerus JA, Green L (2011) How should we define health? Br Med J 343:d4163
35. Intergovernmental Panel on Climate Change (IPCC) (2015) Climate change 2014: synthesis report. Cambridge University Press, Cambridge
36. International Energy Agency (IEA) (2016) Key world energy statistics 2016. IEA/OECD, Paris
37. Kan H, Chen R, Tong S (2012) Ambient air pollution, climate change, and population health in China. Environ Int 42:10–19
38. Kovats RS, Edwards SJ, Charron D et al (2005) Climate variability and campylobacter infection: an international study. Int J Biometeorol 49:207–214

39. Lederbogen F, Haddad L, Meyer-Lindenberg A (2013) Urban social stress—risk factor for mental disorders. The case of schizophrenia. Environ Pollut 183:2–6
40. Le Page M (2016) Invisible killer. New Sci 232(3097):16–17
41. Le Page M (2017) Where there's smoke. New Sci 233(3111):22–23
42. Lelieveld J, Evans JS, Fnais M et al (2015) The contribution of outdoor air pollution sources to premature mortality on a global scale. Nature 525:367–371
43. Li D, Bou-Zeid E (2013) Synergistic interactions between Urban Heat Islands and heat waves: the impact in cities is larger than the sum of its parts. J Appl Meteorol Climatol 52:51–64
44. Marsh M, Coughlan A (2013) China's struggle to clear the air. New Sci 217(2903):8–9
45. Matthews TKR, Wilby RL, Murphy C (2017) Communicating the deadly consequences of global warming for human heat stress. Proc Natl Acad Sci 114:3861–3866
46. McMichael AJ (2013) Globalization, climate change, and human health. N Engl J Med 368(14):1335–1343
47. Mercer 2015. Vienna tops latest quality of living rankings. Accessed on 26 March 2017 at https://www.uk.mercer.com/newsroom/2015-quality-of-living-survey.html.
48. Miller G (2011) Why loneliness is hazardous to your health. Science 331:138–140
49. Monteiro LHA, Chimara HDB, Chaui Berlinck JG (2006) Big cities: shelters for contagious diseases. Ecol Model 197:258–262
50. Mora C, Dousset B, Caldwell IR et al (2017) Global risk of deadly heat. Nat Clim Change 7:501–506. https://doi.org/10.1038/nclimate3322
51. Moriarty P, Honnery D (2014) Reconnecting technological development with human welfare. Futures 55:32–40
52. Moriarty P, Honnery D (2015) Future cities in a warming world. Futures 66:45–53
53. Moriarty P, Wang SJ (2015) Eco-efficiency indicators for urban transport. J Sustain Dev Energy Water Environ Syst 3(2):183–195
54. Pal JS, Eltahir EAB (2015) Future temperature in southwest Asia projected to exceed a threshold for human adaptability. Nat Clim Change 6:197–200. https://doi.org/10.1038/NCLIMATE2833
55. Pecl GT, Araújo MB, Bell JD et al (2017) Biodiversity redistribution under climate change: impacts on ecosystems and human well-being. Science 355:eaai9214. 10 pp
56. Penketh A 2014. Paris mayor announces plans to ban diesel cars from French capital by 2020. The Guardian 8 Dec. Available at http://www.theguardian.com/world/2014/dec/07/paris-mayor-hidalgo-plans-ban-diesel-cars-french-capital-2020.
57. Pickett KE, Wilkinson RG (2010) Inequality: an underacknowledged source of mental illness and distress. Br J Psychiatry 197:426–428
58. Pickett KE, Wilkinson RG (2015) Income inequality and health: a causal review. Soc Sci Med 128:316–326
59. Piontek F, Müller C, Pugh TAM (2014) Multisectoral climate impact hotspots in a warming world. Proc Natl Acad Sci 111:3233–3238
60. Qiu J (2014) Fight against smog ramps up. Nature 506:273–274
61. Riva M, Bambra C, Curtis S, Gauvin L (2011) Collective resources or local social inequalities? Examining the social determinants of mental health in rural areas. Eur J Public Health 21(2):197–203
62. Rohde RA, Muller RA (2015) Air pollution in China: mapping of concentrations and sources. PLoS One 10(8):e0135749. https://doi.org/10.1371/journal.pone.0135749
63. Ruble BA (2012) The Challenges of the 21st-century city. The Wilson Center, Washington, DC, pp 1–4. Available at www.wilsoncenter.org
64. Scrutton J, Holley-Moore G, Bamford S-M (2015) Creating a sustainable 21st century healthcare system. The International Longevity Centre, London. http://www.ey.com/Publication/vwLUAssets/ey-creating-a-sustainable-21st-century-healthcare-system/$FILE/ey-creating-a-sustainable-21st-century-healthcare-system.pdf
65. Sewell AA (2017) Live poor, die young. Nature 545:286–287
66. Sherwood SC, Huber M (2010) An adaptability limit to climate change due to heat stress. Proc Natl Acad Sci 107:9552–9555

67. Sorace C, Hurst W (2015) China's phantom urbanisation and the pathology of ghost cities. J Contemp Asia 46(2):304–322
68. Subramanian M (2016) Delhi's deadly air. Nature 534:166–169
69. Tong S, Prapamontol T, von Schirnding Y (2000) Environmental lead exposure: a public health problem of global dimensions. Bull World Health Organ 9:1068–1077
70. Trivedi J, Sareen H, Dhyani M (2008) Rapid urbanization - its impact on mental health: a South Asian perspective. Indian J Psychiatry 50(3):161–165
71. Underwood E (2017) The polluted brain. Science 355:342–345
72. United Nations (UN) 2012. Changing levels and trends in mortality: the role of patterns of death by cause. Accessed on 21 March 2017 at http://www.un.org/en/development/desa/population/publications/pdf/mortality/Changing%20levels%20and%20trends%20in%20mortality.pdf.
73. United Nations (UN) 2014. World urbanization prospects: the 2014 revision. Accessed on 30 November 2016 at https://esa.un.org/unpd/wup/CD-ROM/. Also the 2011 revision.
74. United Nations (UN) 2015. World population prospects: the 2015 revision. Accessed on 15 December 2015 at http://esa.un.org/unpd/wpp/Download/Standard/Population/
75. United Nations-Habitat (2016) Urbanization and development: emerging futures. In: World cities report 2016. United Nations-Habitat, Nairobi
76. Wikipedia 2017. Air pollution. Accessed on 11 April 2017 at https://en.wikipedia.org/wiki/Air_pollution.
77. Wikipedia 2017. Human development index. Accessed on 26 March 2017 at https://en.wikipedia.org/wiki/Human_Development_Index.
78. Wikipedia 2017. Mercer quality of living survey. Accessed on 26 March 2017 at ttps://en.wikipedia.org/wiki/Mercer_Quality_of_Living_Survey.
79. Wikipedia 2017. Quality of life. Accessed on 26 March 2017 at https://en.wikipedia.org/wiki/Quality_of_life.
80. Wikipedia 2017. Zhuhai. Accessed on 12 April 2017 at https://en.wikipedia.org/wiki/Zhuhai.
81. Wolf T, McGregor G (2013) The development of a heatwave vulnerability index for London, United Kingdom. Weather Clim Extr 1:59. http://dx.doi.org/. https://doi.org/10.1016/j.wace.2013.07.004)
82. Woolf SH, Johnson RE, Phillips RL Jr, Philipsen M (2007) Giving everyone the health of the educated: an examination of whether social change would save more lives than medical advances. Am J Public Health 97(4):679–683
83. World Health Organization (WHO) (2011) Burden of disease from environmental noise: quantification of healthy life years lost in Europe. WHO, Bonn. ISBN: 978 92 890 0229 5
84. World Health Organization (WHO) 2015. World health statistics 2015. Available at http://www.who.int/gho/publications/world_health_statistics/2015/en/
85. World Health Organization (WHO) (2016) World health statistics: monitoring health for the SDGs. WHO, Geneva
86. World Health Organization (WHO) 2017. The top 10 causes of death. Accessed on 18 April 2017 at http://www.who.int/mediacentre/factsheets/fs310/en/.
87. Zhou M, He G, Liu Y et al (2015) The associations between ambient air pollution and adult respiratory mortality in 32 major Chinese cities, 2006–2010. Environ Res 137:278–286

Chapter 3
The Potential for Big data for Urban Sustainability

3.1 Introduction: Traditional Urban Data Collection

Cities have always felt the need for numerical data for their management: the Babylonians conducted the first population censuses almost six millennia ago, mainly to estimate the food needs for their cities [25]. The data were recorded on baked clay tablets, a data storage method that enables them to be read to this day. Even before the advent of modern computers, much detailed information was available about cities, particularly from national censuses. Data typically available included the overall population, its age and sex composition, educational level attained, household size, etc., on a local government level or even on a finer scale. Most post-war censuses also often included information on vehicle ownership, journey to work mode, and job location and category. However, detailed data from the census was often not available to interested users until 1–2 years after the census date. Planners and researchers invariably had to work with obsolete data.

Many cities have also conducted dedicated transport surveys, often as part of a long-range transport plan, although usually not on a regular basis. National level travel surveys, such as those in Australia, the US, the UK and Japan, often have data available on selected cities, such as state capital cities in Australia, and London in the UK. In the case of the US, the regular urban travel data from the National Household Travel Surveys are available only by broad city size category [18].

These travel surveys are not without their problems. The dedicated urban surveys are infrequent, and the more-or-less regular national ones do not have enough detail at the individual city level. There are also problems of data reliability. For many years, the Surveys of Motor Vehicle Usage [4], conducted at roughly 3-year intervals in Australia, included data on total road vehicle petrol usage which was not consistent with national petrol sales data, casting doubt on the travel data as well. Sampling error is also a continuing problem: the relative standard error for some of the Australian data entries exceeds 50%, the result of small sample sizes for subpopulations of people or vehicle types, which greatly reduces their value.

© Springer International Publishing AG, part of Springer Nature 2018
S. J. Wang and P. Moriarty, *Big Data for Urban Sustainability*,
https://doi.org/10.1007/978-3-319-73610-5_3

In the US, there are similar doubts about some of the data collected, with limited comparability between surveys. Apart from sampling errors, non-sampling errors were also found to be a serious problem. The US Federal Highway Administration [18] stated in their report:

'Non-sampling errors in surveys can be attributed to many sources, for example, the inability to obtain information about all persons in the sample; differences in the interpretation of questions; inability or unwillingness of respondents to provide correct information; inability of respondents to recall information; errors made in collecting and processing the data; errors made in estimating values for missing data; and failure to represent all sample households and all persons within sample households (known as under-coverage).'

An increasing problem everywhere is 'survey fatigue' which affects the response rates to surveys [47]. Too many organisations, both government and private, are asking the general population to respond to often time-consuming questionnaires. The resulting low response rates increase the risk of non-representative samples. These sampling and non sampling problems affect all sample surveys to some extent, not only transport ones. The difficulties were amply illustrated by the survey errors in the lead-up to the 2016 US presidential elections. Even census population counts at all levels may be in error because of people who do not wish to have their details recorded, for various reasons. Then there is the problem of the high costs of traditional data collection: the 23rd US census conducted in 2010 cost an estimated US $13 billion [64].

Thus traditional data collection is not without its problems: it can be costly, increasingly difficult to collect, liable to error, and out-of-date. Respondents may knowingly give false information about employment, income, or political attitudes. It is important to stress these shortcomings, as it helps put the discussion in Chap. 4 on the problems of big data in context. Not only will the data needs for urban sustainability grow rapidly, but there are no perfect methods for collecting and processing data. However, probably the greatest problem with traditional recorded data, particularly for urban transport, is its *absence*. Even today, we know very little about the billions of trips humans make in the course of their daily lives. It is accordingly very difficult to formulate good theories to explain past and present urban travel patterns, or energy use because the relevant authorities did not (or could not, for technical or economic reasons) collect the necessary information. The result is that there is not enough data to constrain theories.

3.2   Sustainable Cities Will Need a Rising Volume of Data

The efficient day-to-day operation of modern cities already requires large amounts of data, both in the private and public sectors. (Paradoxically, in the US, at least, some data collection trends are in the opposite direction. According to Fortun et al. [20] 'Recently (particularly since the financial crisis of 2008), in many settings, there have been notable reductions in pollution data collection; too often, insistence on austerity and small government has legitimised closure of monitoring stations and dismissal of technical staff.') With the new US administration in 2017 hostile to environmental legislation, this trend could well continue there. Nevertheless, planning for the future will require even more data, particularly given the constraints imposed by climate change and, possibly, oil depletion. These constraints will lead to unprecedented changes in cities, resulting in past trends being often a poor guide to the future. The reasons for this rising data need are various, as evidenced by the following considerations:

- As we pointed out in Chap. 1, the future will most probably need to see RE play a dominant role as an energy source. In particular, sustainability will eventually require that intermittent sources of RE, mainly wind and solar energy, supply not only most electricity for both cities and elsewhere, but also (after conversion) non-electric energy [40, 41]. Electricity grids once consisted of a few large generating plants—typically 100 MW (MW = megawatt = 10^6 Watt) or larger, sometimes much larger—often running continuously, connected to users by the transmission and distribution system. Varying demand for electricity was met by starting up standby power plants at times of peak demand. In future, particularly with the rise of rooftop photovoltaic (PV) systems, electricity in excess of the instantaneous needs of the individual buildings will be fed into the grid from perhaps millions of sources. Germany is already well down this path. As Chap. 6 will discuss in more detail, electricity grids will need to become—and increasingly are becoming—much smarter, to cope with this rising dual variability in both demand and supply, and the proliferation of energy supply points [9, 43].

- Despite optimistic talk at the 2015 Paris Climate Conference about limiting global temperature rises since pre-industrial times to 2 °C, or an even safer 1.5 °C, some observers doubt that this is now possible, particularly given the need for heavy reliance on the largely untried technology of carbon capture and storage [e.g. 2, 44]. Even if the world's nations make serious efforts to curb global emissions (and the US government in early 2017 does not even recognise the reality of global climate change), it may be difficult to avoid a 3 °C rise by the end of the twenty-first century, if anything like business-as usual economic conditions prevail. Such a rise would move the climate into unchartered climate territory, and already, with a mere 1 °C rise above pre-industrial, we are experiencing an increasing frequency of extreme events, and changes in the world's ecosystems. In short, the world and its regions are moving closer to various tipping points, where abrupt changes can be expected [56]. Averting this increased prospect for instability will require greatly increased *monitoring* of all aspects of the environment by sensors, both in urban and non-urban locations, to predict the approach of such local or global changes, which may be irreversible on human time scales. Such intense monitoring can already be glimpsed in the network of tiny satellites launched by the US company Planet, with 144 already operational in February 2017. This network will enable a complete picture of the entire planet at 3.7 m resolution to be available each day [57].

- It will be increasingly recognised by policy makers that effective action on climate change and the other serious environmental problems will need a *systems* approach. In cities, this will mean recognition of the interconnectedness of presently disparate sectors such as energy, transport, housing, and health. For example, car transport contributes to the Urban Heat Island (UHI) effect in cities, both through its generated waste heat, and its need for large areas of black impermeable pavement surfaces, with further implications for urban health and non-transport air-conditioning energy needs. There will be a need for tradeoffs to minimise overall ecological damage, both for the local urban environment and for the non-urban world, which is heavily influenced by urban consumption and energy use, as discussed in Sect. 1.1.

- Even if the various climate mitigation methods discussed in Chap. 1 were to be adopted more forcefully than has been the case over the past two decades, some climate change would continue, owing to both thermal inertia of the oceans and social inertia of our economic growth-oriented institutions. In cities, this will mean that flooding, together with associated slope instability, heatwaves, and even disease outbreaks, will become more frequent. Nor are these events only a future possibility: heatwaves in Europe since the beginning of this century have killed tens of thousands [42]. In addition to ongoing actions for climate mitigation, it will therefore be necessary to develop and implement policies for climate *adaptation*. To reduce the risk, it is not enough to have real-time and predicted temperatures at the city-wide or even local suburban level. Temperature and humidity levels actually experienced by the population, especially vulnerable members, will also be needed, ideally requiring sensors in each residence or workplace.

- Just considering a single sector, transport, Hubers et al. [34] have pointed out the complexity of the (urban) passenger travel system, with its interaction of many social and technical factors. They have argued that existing data sets are inadequate for the task and that new approaches to data are likely needed for a better understanding that can inform transport policy. Big data will be introduced in urban transport and other sectors because it could prove cheaper than existing methods, a boon for financially-strapped cities. However, demonstrating its clear advantage over more conventional methods will often take time.

- Finally, data needs will rise because of the rising *expectations* of urban residents. They increasingly recognise that it is possible to collect urban environmental information, for example, on an increasingly finer-grained scale. They will no longer be content with city-wide data for air pollution; they will want to know what the pollution levels are on their residential street, and at their workplace, in real time. Such small-area data will be particularly important for the rising number of people with allergies to various types of pollution.

3.3 Big Data, the Internet of Things, and Smart Cities

There are many definitions of 'big data'. Michael Batty [6] mentioned a definition that simply emphasises its sheer size: 'any data that cannot fit into an Excel spreadsheet'. At the Centre for European Nuclear Research, for example, the Large Hadron Collider, used to find the Higg's boson, requires 15 petabytes of computer storage annually [29]. Even so, it discards most of the data points it collects. But according to Tim Harford [29], big data can also be looked at as 'found data'—which he explains as 'the digital exhaust of web searches, credit card payments and mobiles pinging the nearest phone mast.' This explanation directs attention to the fact that vast amounts of potentially useful and relevant data are available as a *byproduct*, unlike the purposeful nature of nuclear research, transport or energy use surveys, or national censuses. Because it was not the focus of attention at the time of collection, its use for some other purpose may only become apparent later. Fortunately, the rapidly falling cost of data storage makes retention of data more feasible. However, not all big data is a byproduct, as the data collected in the focused hunt for the Higgs Boson shows.

The definition of Gudiveda et al. [28] gives further insight. According to these authors, big data can be defined as 'data too large and complex to capture, process and analyze using *current* computing infrastructure'—which again reminds us that the promise of big data lies mainly in the future. They stressed the rapid growth in data production: over 90% of the world's data (as of early 2015) was produced in the two preceding years. As others have also done, they further characterised big data by three V's:

- *volume*—data is already measured in petabytes (2^{50}) and is expanding
- *velocity*—data is produced at very high daily rates
- *variety*—data collected can be heterogeneous.

They added two further V's. First, there may be concerns about the *veracity* of such data, and second, such data can add *value* to business by providing 'counterintuitive insights'. Of course, as already pointed out, concerns about data veracity are also important for traditional data. For Matthew Smith and colleagues [55], the term 'big data' has been mainly used in two different contexts: 'firstly as a technological challenge when dealing with data-intensive domains such as high energy physics, astronomy or Internet search, and secondly as a sociological problem when data about us is collected and mined by companies such as Facebook, Google, mobile phone companies, retail chains and governments.'

From a very different viewpoint, Danah Boyd and Kate Crawford [10] saw big data 'as a cultural, technological, and scholarly phenomenon' with three interconnecting components: technology, analysis, and mythology. While their first two points overlap with other definitions, the third point is a novel claim. They defined big data mythology as 'the widespread belief that large data sets offer a higher form of intelligence and knowledge that can generate insights that were previously impossible, with the aura of truth, objectivity, and accuracy.'

These views on big data can also be compared with the ideas put forward by Victor Mayer-Schönberger and Kenneth Cukier [37] in their book *Big Data*. They stressed that big data 'refers to things that can be done at a large scale that cannot be done at a smaller one', such as enabling new insights. They also gave what they regarded as big data's three essential shifts in our understanding of data and how we use it. First, we can analyse much more data, and in some cases nearly all of it. Second, because we no longer have to rely on sampling, we can work with what they term 'messier' data. As they put it, it's a tradeoff: 'with less error from sampling we can accept more measurement error.' The first two shifts, they argue, lead to the third: a move away from the search for causality, which they regard as a product of the small data era, to a far greater emphasis on correlation alone. Throughout their book, they repeatedly stressed that for many decisions that must be made, 'knowing *what* not *why* is good enough.'

At the heart of the rise of big data are advances in computation, variously called machine learning and data mining, which grew out of the decades' old study of artificial intelligence subfields such as pattern recognition [66]. With machine learning, algorithms—an algorithm is a 'self-contained sequence of actions to be performed' [65]—are constructed which instead of following instructions as in a normal computer program, can *learn* from data inputs and make decisions based on such inputs [19]. Everyday examples include applications such as email filtering and optical character recognition.

One of the reasons to expect that growth of data will be exponential is the anticipated growth of *sensors*. Sensors are devices which collect real-world information, such as ambient temperature, and convert it into a signal, perhaps electrical, which can be interpreted by humans or machines. Their cost has steeply declined in recent years [38]. These sensors could be located anywhere: temperature sensors in buildings; heart rate, blood glucose levels, and other health monitors in or on human bodies; surveillance cameras in buildings and public places (already very common in the UK); door sensors; speed and other sensors in vehicles; condition monitors in machinery and structures, and so on. These 'Internet of Things' (IoT) devices can both generate and share data about the bio-physical environment. According to Weinberg et al. [62], an estimated 16 billion (IoT) devices were already in use worldwide in 2014, and this number could rise to 50 billion IoT devices by 2020.

The term IoT is sometimes used to distinguish it from earlier uses of the Web. When the Web became popular in the mid-1990s, it was mainly a one-way information stream from provider to user/consumer. Web 2.0 allowed for far greater interaction—blogging, social media, file sharing. But IoT devices now 'enable more of the physical and natural world to be integrated into and to become accessible via the Internet' [62].

Smart cities (a related term is intelligent cities) have (or will have) IoT as a vital part of their data network [46]. Just as is the case for big data, there are now many views and definitions of the smart city, not all of them consistent [16, 69]. As Glasmeier and Nebiolo [24] have aptly put it: 'Currently, fragmentary thoughts of what is a smart city reflect rival topics vying for membership in the lexicon of ideas that define smart city status, their purpose, and scope.' One definition was given by Pellicer et al. [48]: 'A smart city is an urban system that uses information and communication technology (ICT) to make both its infrastructure and its public services more interactive, more accessible and more efficient.' They also stated that smart cities have the following six characteristics:

- Smart economy (competitiveness)
- Smart governance (citizen participation)
- Smart people (social and human capital)

- Smart mobility (transport and ICT)
- Smart environment (natural resources)
- Smart living (quality of life).

Paul Budde's [13] list of key components is similar, but he also included the 'three pillars' that smart cities are based on: quality of life, environmental sustainability, and promotion of economic development.

Another related term in common usage is 'green cities'. Green cities are those that are, or aspire to be, environmentally sustainable. They are not necessarily smart cities—many changes important for urban sustainability can be implemented without necessarily being 'smart'. However, in this book, our main aim is to investigate how big data can be used to make cities green; in essence to make them smart green cities.

A number of cities around the world are already describing themselves as 'smart cities'. According to one count, globally there were 22 smart cities in 2013, but this number could increase fourfold to 88 by 2025 [24]. Although big data is already being used in a comprehensive manner in cities such as Rio de Janeiro and New York, and to some extent, in these other smart cities, it is mainly used for urban governance rather than environmental sustainability. By examining the experience of those cities which have implemented one or more aspects of smart mobility or smart environment, we can get an idea of both the potential as well as the challenges that must be overcome if big data is to work for the good of all urban residents [21].

3.4 Present Applications of Big Data/Internet of Things in Cities

Zanella et al. [69] have given a useful summary of how IoT is presently being used to make cities more environmentally sustainable. Although the examples are all drawn from OECD cities, some are already relevant to cities in low- and middle-income countries. The discussion in this section draws on their summary, but also includes other, more recent, information on cities that are actually implementing one or more of these measures, even if the implementation is still in its infancy. It also includes some examples of big data use for urban health and well-being. Section 3.5 will extend this discussion, to show the *future* general potential of big data, especially since, as Glasmeier and Nebiolo [24] have stressed, smart cities have promised much, but so far have delivered few benefits.

3.4.1 Structural Health of Buildings and Other Urban Infrastructure

Sensors allow continuous monitoring of the structural integrity of buildings and other load-bearing structures [63]. Porco et al. [51] have discussed how instrumenting a highway bridge with sensors can allow comparison of actual stresses and strains in the bridge's structural elements at any time with calculated values. Not only does this potentially improve the safety of the structure over its service life, but also allows for control during the construction phase. Visual checks of structures are often inadequate if no information on stresses or displacements is available. Overall, the use of sensors allows maintenance to be more focussed on where it will be most effective.

Further, because actual and calculated stresses can be compared, it opens up the possibility of modifying future designs in the light of this knowledge so that they use less structural materials, and so lower the 'ecological footprint' of urban infrastructure. Essentially, instrumentation can allow full-scale testing of bridges. In cities prone to earth tremors, suitable instrumentation would even make it 'possible to combine vibration and seismic readings to better study and understand the impact of light earthquakes on city buildings' [69].

3.4.2   Waste Management

Several aspects of waste management are important for urban sustainability. Not only is the quantity of waste important, but so are the collection costs and the level of recycling possible. IoT sensors can help in several ways. The city of Boston, USA, has deployed large bins for general trash and recycling that have sensors that can alert officials in real time as to how full they are [33]. This data helps planners decide not only where such bins are needed, but also when they should be emptied, so saving fuel for garbage trucks.

Charith Perera and colleagues [49] have pointed out that waste management, in fact, consists of a number of different processes: 'collection, transport, processing, disposal, managing and monitoring of waste materials.' In future, recycling companies could 'use sensor data to predict and track the amount of waste coming into their plants to be processed, so they can optimise their internal processes.'

3.4.3   Air and Noise Pollution

Most OECD cities (and many cities outside the OECD) today already collect a host of environmental data, such as humidity, temperature, wind speed and rainfall levels. They also collect data on various urban pollutants, including levels of fine particulates, carbon monoxide, oxides of nitrogen and sulphur, as well as a number of other pollutants, such as pollen. In many cases, this data is already remotely sensed.

Although air quality has generally improved in the cities of OECD countries, it is still poor in many cities in industrialising countries. Urban air quality is not a fixed quantity for a given city, but varies by time of day and season, as well as depending on weather conditions. And, of course, it varies greatly from place to place in the city. Air quality instrumentation enables city officials to issue warnings for days of extreme air pollution, usually on a city-wide basis. So far, air quality has been looked at from a purely spatial viewpoint—how it varies for different urban locations. The health effects of various pollutants were then examined, for example, by comparing the health of residents in more polluted areas with those from less polluted areas of the city.

Cartwright [15] has discussed the rise of 'smartphone science' and its possible applications. In the Netherlands, he reports that thousands of residents used an optical device connected to their iPhone cameras to photograph the local sky. 'Within a day, reams of crowd-sourced spectra had stacked up in an online database, ready for analysis.' The data enabled very fine-grained maps of particulate pollution to be generated for this entire small country. Not only does such popular involvement make citizens more aware of atmospheric pollution and its variation, but by using smartphones and a network of sensors, joggers, for example, could now determine the healthiest path to run for their exercise at any given time of the day [50].

Yu Zheng and colleagues [70] at the Microsoft Asia Research Institute have suggested a method to achieve real-time analysis of fine-particle air quality in cities by using a variety of existing data sources, including not only air quality monitoring site readings, but also a variety of other data sources such as weather, traffic flow, 'human mobility', the structure of the road network, and what they term 'points of interest'. Using data from Beijing and Shanghai, their approach proved superior to existing methods. In a later paper, Zheng and co-workers [71] showed how a variety of data sources could be used to predict air quality at a given urban location for the next 48 h.

Noise is now recognised by the World Health Organisation as a cause of illness [42]. Many cities already have policies in place limiting aircraft arrivals and departures at night or limiting heavy trucks from using residential streets at night. Noise limits are also set for individual vehicles. More comprehensive noise monitoring would enable a space–time map of noise to be produced over the instrumented area and to pin-point areas and times with excessive noise levels. But noise monitoring could go much further; by means of sound detection algorithms it could help law enforcement by detecting the noise of glass breaking or gunshots, for example [69]. In the US, police in some areas already have access to sensor networks which detect gunshots [30]. Such detailed information has obvious implications for privacy, as will be discussed in more detail in Chap. 4.

3.4.4 Traffic Congestion/Management, and Parking

Monitoring of traffic flows in cities has been practiced for decades. In the interests of optimising traffic flows, road network managers have long been able to alter speed limits, traffic lane directions and timing of traffic lights, and give real time estimated travel times for road segments.

The SmartSantander project in Santander, a city in northern Spain, has chosen parking monitoring as its first application. It gives dynamic information on general parking availability (as well as on spaces for disabled persons) as a publicly available application [31, 48]. In Boston, 320 downtown parking spots were equipped with occupancy sensors [33]. Not only did the application help motorists find empty parking places, but it also enabled city officials to see how parking policies needed to be changed. Specifically, the maximum time limit on constantly occupied spots was reduced, while for lightly used spots the time limit was raised, allowing more efficient use of available spaces.

3.4.5 Public Transport Information and Promotion

Big data can potentially help existing public transport travellers. For public transport vehicles, many cities, including Melbourne, Australia, now have smart phone apps which can inform travellers (and system operators) of any delays, and give the time of arrival of the next public transport services at their train station, or tram and bus stop. It can also help in planning trips, by giving intending travellers the full range of options available, such as the modes available, and the expected travel time for each choice, including walking time, if any. This information could potentially help in the further shift of urban trips from cars to public and non-motorised transport, as Chap. 5 will detail.

3.4.6 Pedestrian Traffic Counts

In the US, Yin et al. [68] investigated the possibility of using machine learning to assist in counting pedestrian volumes from Google Street Maps of streets in the cities of Buffalo, Boston, and Washington DC. They concluded that their method could detect the presence of a pedestrian with a reasonable degree of accuracy.

3.4.7 Urban Energy Consumption

The urban energy uses we are chiefly concerned with are those for transport, both passenger and freight, and energy use in buildings of all types. Chapter 6 will treat this topic in more detail. Energy savings are often one of the reasons for the other applications discussed here, such as traffic management or waste management. Many cities have already installed smart electricity meters, which supply customers (and operators) with real time information on electricity use, although as explained in Sect. 6.3, simply supplying smart meters has not yet proved enough to motivate householders to reduce their energy consumption; other changes are needed.

The interior environment of buildings is relevant for urban energy consumption, in that the aim is to provide thermal comfort and lighting to occupants while at the same time minimising energy use. Motion sensors, for instance, can detect the presence of occupants, and turn off lights when not needed. They can also even vary lamp intensity, depending on the intensity of ambient lighting. One way to reduce the energy consumption for both building interior and street lighting is to install more efficient lamps. But with the aid of sensors, it is also possible to 'optimize the street lamp intensity according to the time of the day, the weather condition, and the presence of people' [69].

3.4.8 Urban Health and Well-Being

Chapter 7 discusses in detail both the present use of big data for urban health and its potential for both improving health and cutting the rising costs of its provision. Already, many people, both the healthy and the unwell, are measuring a variety of personal data such as blood pressure and blood sugar levels. Monitoring personal health data could become widespread, and encourage the general public to play a larger part in both maintaining healthy lifestyles and illness monitoring.

3.4.9 Urban Governance

Although the examples given above all come from OECD cities, the promotion of smart cities is occurring elsewhere, with India having especially ambitious plans [67]. Probably no city has taken big data further for urban governance than Rio de Janiero, a middle-income megacity, with large numbers of residents living in unplanned settlements, or *favelas*. According to Kitchin [36] this large city has 'a purpose built urban operations centre, staffed by 400 professional workers, which draws together real-time data streams from 30 agencies, including traffic and public transport, municipal and utility services, emergency services, weather feeds, social media and information sent in by the public via phone, Internet and radio.' At the operations centre, analysts use this vast data stream for urban decision-making. 'The result is a new form of highly responsive urban governance in which big data systems are prefiguring and setting the urban agenda and are influencing and controlling how city systems respond and perform.'

Researchers in Beijing are using data from millions of transport smart card swipes on the city's subway system and buses to plot spatial income patterns in this megacity [54]. Although past research had determined the location of low-income residents from other sources of data, the researchers claimed that smart card data was all that was needed. The aim of the research was to 'help the government devise better planning and policies'.

3.5   Future Potential for Big Data in Cities

The above discussion clearly does not exhaust the possibilities for the application of IoT and big data in cities. This section attempts to broadly speculate on how big data could transform cities to a more sustainable future. Social commentator Jeremy Rifkin, in his 2014 book *The Zero Marginal Cost Society* [53], has even argued that the IoT will eventually form an integrated global network, potentially connecting every object with sensors with everyone. 'People, machines, natural resources, production lines, logistics networks, the electricity grid, consumption habits, recycling flows, and virtually every other aspect of economic and social life will be linked via sensors and software to the IoT platform, continually feeding Big Data to every node—businesses, homes, vehicles—moment to moment, in real time'[53]. Whether such a totally interconnected world would be even desirable is another question—security and privacy risks are obvious problems (see Chap. 4).

The limited applications of big data to cities discussed in the previous section have so far had a correspondingly limited impact on solving urban problems. This situation contrasts especially with the use of big data in both the commercial world, and for research. In addition to its use in physics research, big data is now vital for astronomy research, especially for the vast amounts of data from the eight observatories around the world that together form the Event Horizon Telescope which will probe black holes [17]. It has also proved its value for medical research, such as mapping the human genome (although its use for medical research is far more advanced than its use in medical practice.)

Big data is already firmly established on the services provided on the Internet. Book recommendations to customers on Amazon, the online book sellers, were once based on recommendations by human editors. But when Amazon based the advice to customers on correlations between products, this computer-generated advice generated far more sales than the human advice did [37]. Presumably, both the customers and Amazon benefited from the new approach. Similarly, speed recognition, language translation, and word finishing services all rely on the application of big data algorithms.

Already in 2011, Brown et al. [11] were describing its uses in retailing to gain competitive advantage for one retail chain. The company was collecting and analysing sales data from each store, and even each sales unit. It then 'linked this information to suppliers' databases, making it possible to adjust prices in real time, to reorder hot-selling items automatically, and to shift items from store to store easily.' The detailed data also enabled the chain to run real-world experiments to improve profitability. Of course, in the competitive world of retailing, this innovative advantage may not last long.

Data mining in retail can turn up some unexpected but useful results. Data scientists at the US retailer Wal-Mart, combed its customer purchase records to examine how shoppers respond to hurricane warnings. They use shopper behaviour for a hurricane in the same season (Hurricane Charley) to predict what would happen with Hurricane Frances. They found the expected increase in bottled water and flashlight sales, but they also discovered that sales of strawberry Pop-Tarts rose to seven times their normal level [52]. As repeatedly stressed by Mayer-Schönberger and Cukier [37] for a retailer, *why* this totally unexpected finding is the case is not important—what is important is to have adequate in-store supplies of this item.

Another illustration of the predictive potential of big data comes from TV audience research. Shawndra Hill [32] has described how social media data can be used to gauge audience reactions to TV shows. These social media responses are often made during the airing of the show and give audience reactions not only to the show itself, but also to the accompanying advertisements. She found that for 80 popular TV shows broadcast in the US in September-November 2012, Twitter traffic could be used to predict weekly viewer numbers, as validated by the more traditional Nielsen ratings. The detailed data now available can potentially help both the producers of the shows and the advertising sponsors.

Internet-based companies can also run 'A/B tests'. Potential customers can be offered two alternative interface versions of the product, and their responses can help the company quickly decide on the superior version, which may vary by region or country [39].

So far, as the preceding paragraphs show, big data has had far more impact on research and the commercial world than on urban sustainability and well-being. But the ability to run real-world experiments and to receive rapid information on their success or failure is an important feature of big data potential that smart cities will need to pursue. For example, cities can experiment on selected roads with changes to traffic speeds, public transport priority, or parking availability, and quickly assess the impact of the changes. Further, unlike retail stores, big data applications for urban sustainability do not face competition which could erode any temporary advantage, except in so far as cities compete, either nationally or globally.

There are already studies showing how big data could both help us understand how cities function, and how to improve urban policies. One study employed Google Street View to measure incomes at a detailed level in New York city, and found they agreed well with other estimates [23]. It has also been found that on-line customer reviews of San Francisco restaurants on the web site yelp.com was strongly correlated with health department inspection scores for each restaurant. Given limited inspection resources, such ratings could be used to enable better use of inspection staff [23]. In fact, a similar system for Boston found a 30–50% increase in health violations per inspection [3], while a similar program in Chicago yielded a 25% efficiency improvement [45]. This restaurant inspection application, like the use of big data algorithms by Amazon for book suggestions, has no negative consequences if they sometimes get it wrong—all it means is that some inspections are wasted, but far fewer than without the algorithm. Chapter 4 will, however, examine cases where prediction errors can have serious consequences.

The potential applications of big data for city governance are large, but most are in the early stages. For a start, there is the problem with the use of algorithms for decision making, which will be discussed in Sect. 4.1. At present, it is still uncertain whether or not increasing citizen participation and releasing relevant real-time urban data actually leads to improvement in the way cities are run [45]. We also

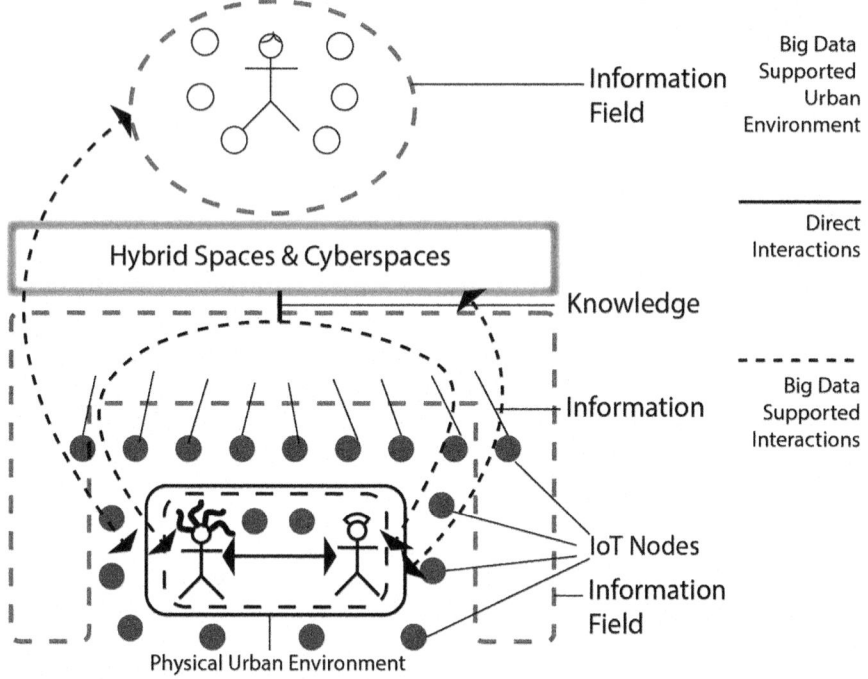

Fig. 3.1 Interactions in a big data supported urban environment

need to know under what circumstances such citizen participation can help make cities more sustainable.

Rob Kitchin [36] has discussed his view of where smart cities are heading, as follows: 'The consequence of the emerging data deluge is that data-informed urbanism is increasingly being complemented and replaced by data-driven urbanism (the mode of production of smart cities) and this is changing how we know, plan and govern cities, both within particular domains (e.g. transport, environment, lighting, waste management, etc.) and across them.' In brief, Kitchin is arguing that big data *will* change cities; the only question is whether these changes can be harnessed for urban sustainability and well-being.

In Fig. 3.1 we show how big data can be used to support an improved urban environment—both physical and social. More interactions will be relevant or backed by big data, and IoT and mobile devices will contribute most of this information gathering process. To avoid the 'information overload' we are presently suffering from (the situation could become worse as IoT progresses), what the users need is not the information (or big data itself) but knowledge which was processed from the big data [8]. The challenge is to provide valuable knowledge to each particular user, through the appropriate media, at the right time and the right place. Therefore, most of the indirect interactions will be mediated by the 'Information Field' which contains ubiquitous AI technologies to convert big data into knowledge [61]. The new technologies could potentially support a broad range of personalised services (for example, Uber, Domino pizza, Airbnb) which could fundamentally change the way people live with limited resources in an already-crowded urban environment.

3.6 Discussion

Does the growing availability of big data, particularly from the Internet of Things, mean that conventional approaches to data collection, and especially analysis methods appropriate to an era of limited data, are now obsolete? Some commentators have thought so, for problems arising from both the bio-physical and the socio-economic domains. In 2008, Chris Anderson, editor of *Wired* magazine, wrote an article that advocated such radical changes: 'Out with every theory of human behavior, from linguistics to sociology. Forget taxonomy, ontology, and psychology. Who knows why people do what they do? The point is they do it, and we can track and measure it with unprecedented fidelity. With enough data, the numbers speak for themselves' [1]. Mayer-Schönberger and Cukier [37], as we have seen (Sect. 3.3), espoused a similar view.

But others are more cautious in their assessment of big data [5, 6, 12, 26, 29, 44, 59]. Batty [6] has a very different slant, arguing that big data will also need big theories. And Keller et al. [35], in an article that looked at what big data can do to improve city living, warned that the traditional approach in the social sciences starts with a *problem* that needs to be solved (or at least studied), such as social inequality, or the transport needs of the urban poor. In contrast, big data approaches start with the *data*, collected from an expanding variety of sources, in the hope that it will be useful in shedding light on various problems. In an era of big data, it is important that we don't become beguiled by 'data fetishism' and fail to focus on the pressing environmental and social problems that all cities face.

We have shown that a number of cities have already adopted elements of the smart city concept to improve the urban environment and urban governance, as well as for urban health. Nevertheless, it is fair to say that so far in no city have these initiatives by themselves resulted in significant reductions in urban energy use or GHG emissions. One possible reason is that the potential for cities to become fully 'intelligent' or 'wired' is still very far from being realised. But future cities will not have a choice as to whether or not to implement mitigation and adaptation policies in the face of climate change; it will be forced on them, most likely by the increasing frequency and severity of extreme weather events in cities and elsewhere. The needed responses will inevitably require greater collection and use of data.

Nevertheless, cities will also need to exercise caution in embracing the smart city idea. Major ICT companies, such as IBM and Cisco, are actively promoting smart cities as a solution to urban problems. Viitanen and Kingston [60] have concluded, based on the experiences of UK cities, that the solutions that private companies promote will 'expand the market for new technology products and services to support 'green growth' with disregard for their wider impacts.' They further warned that what passes for citizen participation or empowerment is often merely consumption of the new technologies.

From a somewhat different perspective, Goodspeed [27] has argued that much of the motivation for smart cities is similar to that for the urban computer models and urban cybernetics developed and promoted decades ago. Urban cybernetics treats the city as a system, with sensors, actuators, and controllers forming a control loop. Critics of this approach have argued that it would not work, because of real world factors such as disagreement about goals among various urban interest groups, city managers pursuing short-term objectives, and the complexity of large cities. More recently, Bettencourt [7] has argued that comprehensive urban planning is 'computationally intractable (i.e. practically impossible) in large cities, regardless of the amounts of data available.' He, therefore, saw a key role of big data in cities as facilitating information flows.

Many of the features of big data analysis are not new. The value of opinion mining has been appreciated at least since the early 2000s. What is new is that there are now cost-effective ways of analysing this heterogeneous data [22]. Nevertheless, big data approaches to urban sustainability problems will not only need to overcome barriers to implementation: privacy, security, and reliability (see Chap. 4). They will have to prove themselves superior to less technologically sophisticated but possibly cheaper approaches. In the transition period, this could prove difficult. We nevertheless think that big data can have a vital role in the uncertain times that lie ahead for the world's cities [42], as they attempt to provide both a high quality of life for all residents while at the same time making sure that none of the planetary limits identified by Steffen et al. [56] are breached. In the chapters that follow we present several case studies of how big data can help make future cities more sustainable.

Nevertheless we need to remember that big data is a necessary but not sufficient condition for future urban sustainability. Some researchers even think that in some cases it may not be necessary. Gina Neff [44] has argued, in the context of the US health care system, that the current enthusiasm for data 'should not overshadow the fact that the United States still needs to provide more basic and preventative health care to more people at a cheaper cost—a problem whose solution we will almost certainly not find in big data.' Similarly, Gudiveda et al. [28] have argued that 'Most problems do not need big data: they need the right data.' Thus big data may not be the solution for all problems society presently faces. We argue here that big data is best used as a vital complement to more traditional approaches in the quest for urban sustainability. It follows that ways must be found to integrate purposeful data from conventional surveys to the far larger data streams that are becoming available that are being collected for other purposes.

So far, this chapter has mainly concentrated on ways in which big data could help OECD countries. However, some believe that Artificial Intelligence (AI) has not only the greatest potential in low-income countries but also the greatest risk for causing harm [14]. As we have already mentioned, this argument once more underlines the point that big data requires supporting social policies if it is to help build a more sustainable and equitable world future.

References

1. Anderson C (2008) The end of theory: will the data deluge makes the scientific method obsolete? Wired. June 23. http://www.wired.com/science/discoveries/magazine/16-07/pb_theory
2. Anderson K (2015) Duality in climate science. Nat Geosci 8:898. https://doi.org/10.1038/ngeo2559
3. Athey S (2017) Beyond prediction: using big data for policy problems. Science 355:483–485
4. Australian Bureau of Statistics (ABS) 2015. Survey of motor vehicle use, Australia, 12 months ended 31 October 2014. Available at http://www.abs.gov.au/AUSSTATS/abs@.nsf/DetailsPage/9208.012%20months%20ended%2031%20October%202014?OpenDocument. Also earlier surveys.
5. Batty M (2013) Big data, smart cities and city planning. Dialogues Hum Geogr 3(3):274–279
6. Batty M, Axhausen KW, Giannotti F et al (2012) Smart cities of the future. Eur Phys J Spec Top 214:481–518
7. Bettencourt LMA (2014) The uses of big data in cities. Big Data 2:12–22. https://doi.org/10.1089/big.2013.0042
8. Bizer C, Boncz P, Brodie ML et al (2011) The meaningful use of big data: four perspectives – four challenges. SIGMOD Record 40(4):56–60
9. Blumsack S, Fernandez A (2012) Ready or not, here comes the smart grid! Energy 37:61–68
10. Boyd D, Crawford K (2012) Critical questions for big data: provocations for a cultural, technological, and scholarly phenomenon. Inf Commun Soc 15(5):662–679
11. Brown B, Chui M, Manyika J (2011) Are you ready for the era of 'big data'? McKinsey Quart 24:24–35
12. Brownell B (2014) The new look of analytics. Research World 2014:26–31
13. Budde P (2014) Smart cities of tomorrow. Chapter 12. In: Rassia ST, Pardalos PM (eds) Cities for smart environmental and energy futures, Energy systems. Springer, Berlin. https://doi.org/10.1007/978-3-642-37661-0_2
14. Butler D (2017) AI summit aims to help world's poorest. Nature 546:196. Accessed on 8 June 2017 at https://www.nature.com/articles/n-12339880
15. Cartwright J (2016) Smartphone science. Nature 531:669–671
16. Chourabi H, Gil-Garcia JR, Pardo TA et al (2012) Understanding smart cities: an integrative framework. 45th Hawaii International Conference on System Sciences. https://doi.org/10.1109/HICSS.2012.615
17. Clery D (2017) Global telescope gears up to image black holes. Science 355:893–894
18. Federal Highway Administration 2011. Summary of travel trends: 2009 National Household Travel Survey. Report no. FHWA-PL-ll-022
19. Fernández MR, García AC, Alonso IG, Casanova EZ (2015) Using the big data generated by the smart home to improve energy efficiency management. In: Energy efficiency. Springer, Cham. https://doi.org/10.1007/s12053-015-9361-3
20. Fortun K, Poirier L, Morgan A, Costelloe-Kuehn B, Fortun M (2016) Pushback: critical data designers and pollution politics. Big Data Soc 3:1–14. https://doi.org/10.1177/2053951716668903
21. Galdon-Clavell G (2013) (Not so) smart cities?: the drivers, impact and risks of surveillance enabled smart environments. Sci Public Policy 40:717–723
22. Gandomi A, Haider M (2015) Beyond the hype: big data concepts, methods, and analytics. Int J Inf Manag 35:137–144
23. Glaeser EL, Kominers SD, Luca M et al (2016) Big data and big cities: the promises and limitations of improved measures of urban life. Econ Inq 56:114–137. https://doi.org/10.1111/ecin.12364
24. Glasmeier AK, Nebiolo M (2016) Thinking about smart cities: the travels of a policy idea that promises a great deal, but so far has delivered modest results. Sustainability 8:1122. https://doi.org/10.3390/su8111122
25. Gomez CG (2013) Great moments in statistics: ancient censuses. Significance 10:21

26. González-Bailón S (2013) Social science in the era of big data. Policy Internet 5(2):147–160
27. Goodspeed R (2015) Smart cities: moving beyond urban cybernetics to tackle wicked problems. Cambridge J Regions Econ Soc 8:79–92. https://doi.org/10.1093/cjres/rsu013
28. Gudivada VN, Baeza-Yates R, Raghavan VV (2015) Big data: promises and problems. Computer 48:20–23
29. Harford T (2014) Big data: are we making a big mistake? Significance 11:14–19
30. Hvistendahl M (2016) Crime forecasters. Science 353:1484–1487
31. Haubensak O (2011) Smart cities and Internet of things. In: Michahelles F (ed) Business aspects of the Internet of things, Seminar of advanced topics, FS2011. ETH, Zurich
32. Hill S (2014) TV audience measurement with big data. Big Data 2:76–86
33. Hodson H (2015) A city of numbers. New Sci 225(3005):22–23
34. Hubers C, Lyons G, Birtchnell T (2011) The unusual suspects: the impacts of non-transport technologies on social practices and travel demand. In: 43rd Universities Transport Study Group Conference, 5th-7th January 2011, Milton Keynes, UK
35. Keller SA, Koonin SE, Shipp S (2012) Big data and city living – what can it do for us? Significance 9:4–7
36. Kitchin R (2016) The ethics of smart cities and urban science. Philos Trans R Soc A 374:20160115. https://doi.org/10.1098/rsta.2016.0115
37. Mayer-Schönberger V, Cukier K (2014) Big data. Mariner Books, Boston, New York, NY
38. McGrath MJ, Ni Scanaill C (2015) Sensor technologies: healthcare, wellness and environmental applications. Apress Open/Springer, New York, NY
39. Mitchell T, Brynjolfsson E (2017) Track how technology is transforming work. Nature 544:290–292
40. Moriarty P, Honnery D (2011) Is there an optimum level for renewable energy? Energy Policy 39:2748–2753
41. Moriarty P, Honnery D (2011) Rise and fall of the carbon civilisation. Springer, London
42. Moriarty P, Honnery D (2015) Future cities in a warming world. Futures 66:45–53
43. Moriarty P, Honnery D (2016) Can renewable energy power the future? Energy Policy 93:3–7
44. Neff G (2013) Why big data won't cure us. Big Data 1(3):117–123
45. Noveck BS (2017) Five hacks for digital democracy. Nature 544:287–289
46. O'Grady M, O'Hare G (2012) How smart is your city? Science 335:1581–1582
47. Olson CA (2014) Survey burden, response rates, and the tragedy of the commons. J Contin Educ Health Prof 34(2):93–95
48. Pellicer S, Santa G, Bleda AL et al (2013) A global perspective of smart cities: a survey. In: Seventh International Conference on Innovative Mobile and Internet Services in Ubiquitous Computing. https://doi.org/10.1109/IMIS.2013.79
49. Perera C, Zaslavsky A, Christen P, Georgakopoulos D (2014) Sensing as a service model for smart cities supported by Internet of Things. Trans Emerg Telecommun Technol 25:81–93
50. Pilla F, King E, Broderick B et al 2012. Real-time measurement of inhabitants exposure to noise and air pollutants with a network of sensors. Accessed on 11 January 2016 at http://senseable.mit.edu/papers/pdf/20120825_Pilla_etal_RealtimeAssessment_SESEH.pdf
51. Porco F, Fiore A, Porco G et al (2013) Monitoring and safety for prestressed bridge girders by SOFO sensors. J Civil Struct Health Monitor 3:3–18
52. Provost F, Fawcett T (2013) Data science and its relationship to big data and data-driven decision making. Big Data 1(1):51–59
53. Rifkin J (2014) The zero marginal cost society: the Internet of Things, the collaborative commons, and the eclipse of capitalism. Palgrave Macmillan, New York, NY
54. Rutkin A (2015) Subway betrays status. New Sci 225(3009):22
55. Smith M, Szongott C, Henne B et al 2012. Big data privacy issues in public social media. 6th IEEE DEST Conference. Accessed on 27 March 2017 at http://ieeexplore.ieee.org.ezproxy.lib.monash.edu.au/stamp/stamp.jsp?arnumber=6227909.
56. Steffen W, Richardson K, Rockström J, Cornell SE, Fetzer I, Bennett EM et al (2015) Planetary boundaries: guiding human development on a changing planet. Science 347(6223):1259855. (10 pp)

57. Strauss M (2017) Planet Earth to get a daily selfie. Science 355(6327):782–783
58. Su X, Shao G, Vause J et al (2013) An integrated system for urban environmental monitoring and management based on the Environmental Internet of Things. Int J Sustain Dev World Ecol 20(3):205–209. https://doi.org/10.1080/13504509.2013.782580
59. Taleb N 2013. Beware the big errors of 'big data'. Wired blog. Accessed on 15 December 2015 at http://www.wired.com/2013/02/big-data-means-big-errors-people/).
60. Viitanen J, Kingston R (2014) Smart cities and green growth: outsourcing democratic and environmental resilience to the global technology sector. Environ Plan A 46(4):803–819
61. Wang SJ (2013) Fields Interaction Design (FID): the answer to ubiquitous computing supported environments in the post-information age. Paramus, NJ, Homa & Sekey
62. Weinberg BD, Milne GR, Andonova YG et al (2015) Internet of things: convenience vs. privacy and secrecy. Bus Horiz 58:615–624
63. Wikipedia 2016. Internet of things. Accessed on 21 Jan 2016 at https://en.wikipedia.org/wiki/Internet_of_Things.
64. Wikipedia 2016. 2010 United States Census. Accessed on 20 Jan 2016 at https://en.wikipedia.org/wiki/2010_United_States_Census.
65. Wikipedia 2017. Algorithm. Accessed on 4 April 2017 at https://en.wikipedia.org/wiki/Algorithm.
66. Wikipedia 2017. Machine learning. Accessed on 4 April 2017 at https://en.wikipedia.org/wiki/Machine_learning.
67. Wikipedia 2017. Smart city. Accessed on 4 April 2017 at https://en.wikipedia.org/wiki/Smart_city.
68. Yin L, Cheng Q, Wang Z (2015) 'Big data' for pedestrian volume: exploring the use of Google Street View images for pedestrian counts. Appl Geogr 63:337–345
69. Zanella A, Bui N, Castellani A et al (2014) Internet of things for smart cities. IEEE Int Things J 1(1):22–32
70. Zheng Y, Liu F, Hsieh H-P (2013) U-Air: when urban air quality inference meets big data. In: KDD'13, August 11–14, Chicago, Illinois, USA. Available at https://pdfs.semanticscholar.org/5ff3/b3cb15c3eb95484ab5b9d5e63b1858521b3a.pdf
71. Zheng Y, Yi X, Li M et al (2015) Forecasting fine-grained air quality based on big data. In: KDD '15, August 11-14, Sydney, NSW, Australia. https://doi.org/10.1145/2783258.2788573

Chapter 4
Barriers to the Implementation of Big Data

4.1 Introduction

Before application of big data and IoT can realise its potential for urban sustainability, a number of barriers to its use must be overcome. These include the challenges of maintaining data privacy and security, and ensuring the reliability of any data collected or used. And although big data is already producing financial returns in some areas of application, in others it still has to prove its worth economically. Because big data is presently being employed more extensively in areas other than smart cities, both these challenges, as well as the benefits, are presently more salient in these areas. Accordingly, we include some examples from sectors not directly concerned with urban sustainability, such as marketing and retail sales.

4.2 Privacy Problems

The problem of privacy predates computers, let alone big data; think of the widespread surveillance of the population in Stalinist USSR or Nazi Germany in the 1930s and 1940s—or East Germany until the late 1980s. Nevertheless, worries have intensified over recent decades, particularly with the rise of the Internet and social networking. A 2017 *New Scientist* article [2] provocatively entitled '*Your entire internet history is now for sale*', sums up the privacy problem in the internet age. Privacy, traditionally defined as the ability to keep at least some information about our thoughts, activities, and finances secret from others, is seen as essential for humans, for our functioning as citizens [39, 63]. As Oboler et al. [42] have pointed out, at least some personal information should not be generally available. It is for this reason that wiretapping is legally restricted, and access to our medical records is protected. But in the US, the new Trump administration has rescinded a bill that

S. J. Wang and P. Moriarty, *Big Data for Urban Sustainability*,
https://doi.org/10.1007/978-3-319-73610-5_4

would have prevented Internet Service Providers from selling the data they collect about all users to other parties, without explicit permission from Internet users [2].

Nevertheless, we have long been willing to share personal information with the taxation office, for example, because without any details of our finances and personal information such as number of dependants, there would be few alternatives to everyone paying the same tax, regardless of circumstances. Similarly, physicians and psychotherapists need detailed information about their patients if they are to treat their ailments effectively. Undoubted societal benefits can occur too, as when electronic health records from an entire population are combined to give new medical insights, such as novel drug interactions. But as Acquisti et al. [1] have noted: 'On the other hand, the potential for personal data to be abused—for economic and social discrimination, hidden influence and manipulation, coercion, or censorship— is alarming.' They further warned that: 'Sharing more personal data does not necessarily always translate into more progress, efficiency, or equality.' The risks are especially great when body sensors become more common, as these will inevitably generate personal health data of significant commercial interest—to life insurers, for example. Matthew Smith and colleagues [54] have discussed an interesting example from social media. What if a friend uploads a picture of a person participating in a volleyball game, and uploads it on a social media site? Insurance companies could use this picture to deny an insurance claim or to raise rates for that person. Some 300 million pictures are uploaded daily to Facebook [27].

Even data that has in theory been anonymised can violate privacy expectations. Recent research showed that by data mining of public records, anonymised profiles could be correctly matched to individuals with 84–97% accuracy [15, 40]. So, in an era of big data, 'de-anonymisation' is possible. What does consent to use personal data anonymously mean if in an era of big data it is no longer really anonymous? And will the risk of such privacy violations dry up this source of data?

With the advent of big data, concerns about privacy have grown, but many observers feel that the traditional methods for preserving privacy will no longer work, that we need to think about privacy in a new way. Jens-Erik Mai [34] has argued that we need to shift the emphasis from whether individual consent was obtained for personal data collection 'to concerns about the new insights that others can generate based on the data already available.' These new ways of using the data may not be fully apparent to anyone at the time the data was collected, either the individuals whose personal data is being collected or even the data miners themselves. It is probable that those who are inexperienced in using the Internet will be adversely affected the most. They may not appreciate the uses to which their personal data could be put, compared with more experienced or better-educated users [29].

Viktor Mayer-Schönberger and Kenneth Cukier [36], when discussing the dark side of big data, have stressed another problem, which they called 'propensity'. Although not strictly a privacy problem, it will be discussed here because it directly concerns data collected on individuals and groups of individuals. It manifests itself in the present debate on 'predictive policing'. This approach is not new, as it has been discussed since the 1930s. Mara Hvistendahl [26] has described how it works. In 2016, plans were announced for a predictive policing trial in Homewood, a predominately black suburb in Pittsburgh, using algorithms developed at Carnegie Mellon university. Similar programs are already in operation in other cities in the US. In Los Angeles, police are provided with computer-generated maps on their patrol car laptops showing 'hot spots' of past crimes, so that patrols can be increased in these areas. Proponents have claimed that it can not only reduce crime but help reduce racial bias in policing, by having decisions on more objective grounds.

Opponents on the other hand, in addition to pointing out the privacy risks, have claimed that it could increase racial bias by moving such bias into the supposedly objective algorithms used to analyse the input data. Kate Crawford and Ryan Calo [14] have documented examples of this hidden bias, citing a 2016 study which found that 'the proprietary algorithms widely used by judges to help determine the risk of reoffending are almost twice as likely to mistakenly flag black defendants than white defendants.' Aaron Shapiro [53] has given a further reason for distrusting the algorithms used in present predictive policing systems: the training algorithms omit 'certain crime types and data sources', which could increase racial bias. In a simulation experiment using the PredPol algorithm for Oakland, California, researchers found that it would mainly direct police officers to Latino and African American precincts, despite an official survey that showed that 'drug use is dispersed evenly across all of Oakland's neighbourhoods.' It does not have to be this way: algorithms can be trained to make fairer decisions [48].

Is predictive policing successful in its declared aim of reducing the urban crime rate? A report from the RAND Corporation found that the advantage over 'best practice' conventional policing was at best incremental. Leaving aside this argument, there is a deeper issue highlighted by Mayer-Schönberger and Cukier [36]. They have pointed out that our legal systems are based on the *presumption of innocence*, and that people are not to be punished for crimes they might commit in the future, or in their words 'penalties based on propensities'. There is a danger that with a general reliance on big data for prediction in the criminal justice system (as, for example, in determining whether parole should be granted) that civil rights could be ignored. Already, judges in some US states make their sentencing decisions based on algorithms that determine the probability that the accused will re-offend [48].

This is particularly likely to be a problem if it is found that the use of big data algorithms is successful in reducing re-offending among parolees, or in lessening crime in certain areas. And what if the use of big data enables the total prison population to be significantly reduced? Alternatively, if it proves effective in reducing terrorist attacks? Will we then still be as concerned about personal guilt or innocence?

It is not only in policing that the danger of discrimination exists. In a manner similar to predictive policing, employers who had access to the relevant records might vet potential employees on the basis of predicted probability of work absences or contracting a disease [23]. Similar discrimination practices could occur for credit risk rating or granting mortgages. A person may be an excellent credit risk, but if they have some characteristics common to people who are statistically bad risks, that person could be grouped in this category. Again, its effectiveness in reducing costs could be used as a justification.

The privacy challenges may be less serious for much of the big data relevant for urban sustainability than those for more commercial purposes. For instance, it is hard to think of any privacy problems arising from examples discussed in Sect. 2.4—instrumenting bridges (or buildings and machinery) with sensors for optimal maintenance and design, or measuring particulate pollution with smartphones. But unless special precautions are taken, the connection of millions or even billions of sensors to the internet will inevitably increase the risks to privacy [67].

Further, many sensors will be located in homes or on or even in the bodies of individuals. They will presumably only be installed with the agreement of the owners, and, ideally at least, the owners can choose how the data is used, and what level of privacy they want. If the owners feel their data is being misused, they could simply disable the sensors. One possible approach to implementation is that owners can in effect sell the data available from these sensors to data companies [45]. The problem is, of course, that the owners—or anyone else for that matter—will often have little idea as to what uses the data will eventually be put. Disabling sensors or stopping data sales may be too late if the data which is now causing concern has already been collected, sold and used.

However, even this limited control over the data does not apply to the millions of closed-circuit TV (CCTV) cameras installed in a large variety of locations, including workplaces, shopping malls, car parks, public transport, or public areas [64]. The reasons for installation were similarly varied: for crime prevention, for monitoring employees and for home and school security reasons. (Many were also used for remote monitoring of industrial processes in, for example, oil refineries, with no implications for privacy.) The only way the individual can maintain full privacy is by completely avoiding the areas under surveillance, which is becoming increasingly impractical, given that there were over 350 million CCTV cameras installed worldwide by 2016.

How important do the general public regard the privacy issue? When we consider the personal information that is shared on social networking sites, particularly by younger people, it might be thought that the answer is: not very much. As Utz and Krämer [60] have pointed out, there appears to be a 'privacy paradox' with users of social networking sites: in surveys, respondents express concern about privacy, but on the other hand, their profiles often give details of their personal lives. Nevertheless, most of the respondents to their survey reported that they protected at least some parts of their profiles, which suggests that they are at least aware of the privacy problem, and do want to place some limits on what personal details they wish to share with others.

In another study, Stutzman and colleagues [57] followed use by over 5000 users on the Carnegie-Mellon University Facebook network in 2005 through to 2011. They found that over the period, users did consciously try to improve their privacy by restricting the personal information they made publicly available. But that trend was partly reversed by changes to Facebook privacy settings. In fact Facebook privacy policies have been subject to numerous criticisms [65], which fortunately have led to policy changes.

So far, we have argued that big data will generally heighten privacy risks. Nevertheless, a contrary view is that, in general, such applications in smart cities will produce a mixture of both gains and losses to privacy [43]. Cavoukian and Jonas [13] in their report *Privacy by Design* also have claimed that privacy concerns can be effectively addressed if they are embedded in the design of the application.

4.3 Security Problems

Security, rather than privacy, could be the main problem with some types of big data. In the following paragraphs, we discuss the security challenges to smart cities in general and big data applications in various sectors such as health or urban infrastructure.

Sam Wong [66] believed that the smarter cities become, the more vulnerable they will be to cyber attacks, and that, further, smart cities are not doing enough to protect themselves. He argued that: 'Unlike companies, which have a unified leadership and policies, cities are fractured into public and private organisations, making them much harder to defend against cyberattacks.' He believed that security is a particular problem for large cities, because of their complexity. Further, it is probable that the smarter the city—the more interconnected the various city functions are—the greater the risk of security breaches.

As Weinberg et al. [63] have pointed out, a security breach for IoT could in some cases be even more serious than existing threats in other areas such as credit card fraud, since physical objects will be increasingly connected to the Internet. We have already pointed out that there are few privacy concerns with IoT data for the structural condition of bridges. On the other hand, the security (safety) implications could be large if the data were hacked or misrepresented. Imagine if the power supply system for a city was hacked by either individuals—or foreign governments as part of their covert cyber warfare. There are already several cases where computer systems were believed to have been compromised by foreign powers. While the origin of the cyber attack is still debated, it is a fact that an introduced computer worm was able to change the physical operation of uranium enrichment centrifuges in Iran, destroying hundreds of them [51] It also disrupted industrial processes in several other industrialising countries [30]. The consequences of such an attack on the operation of a nuclear power station close to a large urban centre could be catastrophic.

Another important concern is the security of vehicles, which are becoming increasingly 'intelligent', with a few fully automated vehicles—such as Google cars—already on the roads. In a recent test, security researchers were able to remotely switch off the engine of a Jeep vehicle while it was travelling on the road [7]. A nightmare security scenario with automated vehicles could involve a driverless car bomb [8]. As will be shown in Chap. 5, automated vehicles face other difficulties as well, but the security implications alone may be enough to abandon this development.

We have already highlighted the privacy problems associated with health records, but there are potential security problems as well. What if the sensors for IoT are not merely connected to things, but to people? For instance, it is not difficult to imagine the physical safety risks that would result if sensitive health devices like pacemakers, connected through the Internet to healthcare specialists for monitoring, were hacked. In early 2017, the US Food and Drug Administration warned that unauthorised users could potentially hack RF-controlled cardiac devices such as pacemakers, and change the heart pacing or deplete the battery [50].

However, as is the case for privacy, in some ways the security situation with smart city IoT is better than for commercially important data. Unlike, for instance, theft of credit card data, or confidential corporate information, at least there is little private monetary gain to be had from hacking into city air pollution or noise level data. Nevertheless, given that a number of governments probably engage (or are planning to engage) in cyber warfare [51], the security of much city data—and physical devices connected to the Internet—is vital.

The various applications of big data discussed in the following chapters must have security built into the systems [33]. In many applications to data, security considerations were not given the importance they deserved; in the rush to get new commercial products to the marketplace, security was often treated as merely an add-on [44]. The risks were starkly illustrated in the May 2017 cyberattack, one of several recent ones, which infected 'more than 200,000 computers in 150 countries' [4]. The hackers demanded a ransom from each victim to fix the problem. The UK National Health Service was among the many victims, resulting in doctors being unable to bring up their patients' medical records. The success of the attack was the result of poor security measures, in turn, the consequence of misguided attempts to save money. This will not do if big data is to have a future in the various aspects of urban sustainability. Security breaches tend to be well-publicised, and serious ones could destroy public support for their application in healthcare or smart grids, for example.

The risk of security breaches is greater in industrialising countries. Nir Kshetri [30] has referred to the 'hollow diffusion' of the Internet and its applications such as e-commerce. He explained what he meant by this term as follows: 'Many organizations digitizing their activities lack organizational, technological and human resources, and other fundamental ingredients needed to secure their system [...]'. Particularly in countries with pervasive corruption, it would be a mistake to rely on the Internet of sensitive applications, including health data. But there are positive aspects for security as well: in the OECD countries, at least, big data techniques have been used by the banking industry to spot fraud [55].

4.4 Reliability Problems

Although Chris Anderson [6] has claimed that the 'numbers speak for themselves', we now know better, partly as a result of the Google Flu Trends experience. Google researchers claimed in 2009 that 'by analysing flu-related queries coming into its search engine, it can accurately detect flu outbreaks quicker than the Centers for Disease Control and Prevention' [16]. However, a 2014 paper in *Science* by David Lazer and his colleagues [32] showed that the early predictive successes of Google Flu Trends were not sustained. The main problem was with the algorithm the Google search engine used to analyse search queries, which is continuously modified and improved. 'As a result, the data that were collected at different times had different purposes, causes, and interpretations' [16].

Reliability problems can arise from either deliberate manipulation of the data, inadvertent errors, or a mix of both. One example of the latter is 'MyShake', a seismology app for smartphones being developed at the University of California for use by citizen scientists. The developers are having to refine the app 'so that it can distinguish between an actual seismic event and when a user is shaking the phone.' Nevertheless, as one seismologist commented: 'The prospect that seismic data in large earthquakes can be obtained from consumer electronics is potentially transformative' [12]. Researchers are confident that the reliability problems of this and other applications can be solved. As stressed earlier in Sect. 3.1, reliability is a serious problem for conventional survey data as well, caused by faulty recall or, as with big data, deliberate misrepresentation.

David Lazer and others [32] have offered another example of the deliberate manipulation of data, this time from social media, for financial or political gains. Socia media have been used to attempt to influence the stock market, and the political process. As the authors stressed, as we progressively monitor more of people's behaviour using media like Twitter or Facebook, the greater the incentive for interested parties to manipulate social media responses. Manipulation can occur in other ways. Concerning big data's use in the allocation of health inspections (see Sect. 3.5), restaurant owners could learn how the algorithm is used to make inspection visits. If, for example, their restaurant is in a low-visit category, they might relax their health standards. Susan Athey [5] has discussed an example from the timber industry in British Columbia, where the possibility of bidding manipulation for government-owned forests led the provincial government to sacrifice some of the predictive power of their algorithms to hedge against this possibility.

In a 2016 study, Wang et al. [61], examined the reliability of big data for political events culled from news media reports, and found major problems with reliability and validity. They looked at big data records of protest events in Venezuela during 2013–2014 and found that only about 21% were about a real protest event. They concluded that automated event coding showed great promise, but still had a long way to go.

Having lots of data is no guarantee against bias. A celebrated example, well before the advent of digital computers, is the 1936 survey of voters' intentions in the upcoming US presidential elections by the US-based *Literary Digest*. Despite a polling sample of more than two million, their prediction proved hopelessly wrong: instead of Republican candidate Alf Landon receiving the predicted 57.1% of the votes, he actually received only 36.5%. Franklin Delano Roosevelt, the incumbent Democrat, won by a landslide. The trouble was that the poll was not representative of the US voting population. It used phone calls and car registrations for polling, which biased the sample toward higher income groups, who were less likely to be impressed by Roosevelt's progressive 'New Deal' policies [46].

A vast and growing literature has examined the data trail from social media such as Facebook and Twitter, looking at such topics as social networks, how people interact with each other, and how news—and misinformation—is propagated. But even if, for example, *all* tweets for a given period are examined, it only gives us information about users of Twitter, not the general public. Moreover, as Eszter Hargittai [25] has documented for the US, survey results show that use of social media varies across the general population, depending on such factors as their age, gender, ethnicity, and income. It also obviously varies with their ability to access the Internet and their familiarity with using it. For instance, when investigating social support in the community, those who actively use Facebook are already more likely to enjoy social support, thus biasing the conclusions drawn. The demographics can even be different for the different social networking sites. And as David Lazer [31] has stressed, the data exhaust we do collect is only a tiny fraction of the data that could conceivably be obtained. Further, it is not necessarily what would have been collected in a purposeful survey to address a given problem.

A different reliability problem occurs even if the data is accurate, and arises from the *analysis* of the data. As Nassim Taleb [58], the originator of the concept of 'Black Swans' has stressed, 'big data means anyone can find fake statistical relationships, since the spurious rises to the surface. This is because in large data sets, large deviations are vastly more attributable to variance (or noise) than to information (or signal)'. He accordingly warned that any large deviations found during data analysis were likely to be bogus.

Put another way, there is increased risk of 'Type I errors'—mistakenly asserting the presence of a relationship when there isn't one [52]. The huge sample sizes made possible by big data can make 'insignificant findings seem meaningful because they achieve conventional thresholds of statistical significance'. In a similar vein, Ganguly et al. [19] have argued for using big data and data mining to improve understanding of extreme climate events. But they also caution that: 'Physically based relationships (where available) and conceptual understanding (where appropriate) are needed to guide methods development and interpretation of results.'

Nevertheless, with big data, we do not need perfect data. The IoT will connect vast numbers of sensors used for a variety of purposes. Obviously, not all of these sensors, which already number in the billions, will always function as planned. There will be many false readings, or no readings at all if their power source fails. Even if these spurious readings are included in analyses, it will cause little error. Mayer-Schönberger and Cukier [36] have given an example of how this works. We could use one sensor to measure the temperature of a whole vineyard. Clearly, if this sensor is malfunctioning, we will draw the wrong conclusion. But if we install many sensors, they can be both cheaper and less sophisticated. This may increase the chances that some will malfunction, but overall, the combined data will give us a good picture of how temperature varies in the vineyard—now both spatially and temporarily. Only if there are reasons to believe that the sensors are systematically biased will a misleading result be obtained.

4.5 Technical problems

Big data faces several other problems hindering its more widespread adoption in urban management, such as supplying the energy for remote access to IoT sensors, and integrating the various data streams from many different sources. These and other issues are already well-recognised by big data researchers [18, 22]. A survey of IT managers [59] revealed that they presently lack data storage, access technology, computational power and even the personnel needed for extracting value from big data. If these represent serious problems in OECD economies, they are even more of a challenge in many low income countries, where there is simply a lack of the necessary trained personnel and IT infrastructure to support big data applications.

Richard Haight and colleagues [24] have discussed the problems of powering the billions of sensors at what they call the 'edge of the Internet'. Various possibilities for powering these devices include harvesting ambient energy, which could use kinetic, solar, thermal or wind energy. The total power requirement of the sensor depends upon whether it must operate continuously or periodically. Depending also on the sensor function, power needs will usually range over several orders of magnitude, from ~100 µW to ~10 mW. The power is needed for sensing, processing this data and storing it, and transmitting it to a receiver. Of course, for sensors installed in buildings, wiring all such devices to the power supply is an option, but would be far too expensive, given their numbers. For sensors inside buildings, PV cells responding to ambient light are an obvious choice. But for outdoor sensors, energy storage will be needed to ensure a continuous and reliable power supply, perhaps in the form of small lithium batteries. Storage will further add to the cost of the IoT, and restrict it to applications where the benefits clearly outweigh the costs.

One answer to the problem of power sources for sensors is to use the sensors in mobile phones, where batteries are kept charged by the owner. There were an estimated 4.77 billion phones—and 8.24 billion mobile phone connections—in the world in 2017, and a growing share of these phones are smart phones [56]. Not only are ownership levels still growing, but smart phones are becoming ever more sophisticated. We have already discussed their application for pollution measurement and tremor sensing in earthquake-prone areas. Their use could be extended to monitor other local environmental conditions by volunteer 'citizen scientists'.

Even Moore's 'law', which states that 'the number of transistors on a microprocessor chip will double every 2 years or so' may be reaching the end of its life [62]. As miniaturisation progresses, it will eventually reach a stage where quantum effects render transistors unreliable. Researchers are thus looking at new approaches to keep the IT revolution progressing, which will be necessary for processing the vast amounts of data that will be collected.

The data streams that big data use are heterogeneous in both type and source. Some will come from sensors measuring physical properties of the ambient environment: temperature or concentration levels of various urban pollutants. Other data will come from far more subjective sources such as the text and images of social media, where interpretation of what is meant is subjective. The coverage of the population at large may be incomplete, and the sample while very large, may not be representative. Is the data valid, or is much of it spam, or is it being consciously manipulated?

4.6 Cost Problems

All the challenges the application of big data faces: privacy, security, reliability, and any remaining technical problems, are likely to be reflected in the costs for its application. Once the use of big data is widespread, and business and the general public appreciate this, attempts at manipulation for monetary or political gain could become common, and reliability will suffer. In the extreme case, much of the data based on opinions as reflected in social media could become suspect; people could deliberately give misleading information as a reaction to their personal information being used to profit others. We could see a repeat of what has happened with security for networked computers, with a never-ending contest between hackers and security systems. In this case, data miners will have to extract value from potentially suspect data. Of course, much data won't be subject to manipulation—a person cannot fake a purchase at a retail store or fake travel on a certain stretch of an urban freeway at a certain time, and usually have little to gain by trying to manipulate their own health data, such as blood sugar levels. Obtaining adequate levels of security, whether against data theft of private medical data or corporate data, or against malicious hacking or even cyber warfare could both limit possible applications of big data and increase costs.

In the transition phase to the widespread use of big data for sustainable cities, the economics is likely to look doubtful. Also, the various challenges discussed above, particularly security and reliability, are likely to raise the cost of using big data. Natalie Allen [3] has even claimed that the IT security problems presently facing smart cities could make them not only riskier, as already discussed, but also more expensive to operate than conventionally-run cities. This questioning of the economics of new technology happened in an earlier phase of computerisation. Brynjolfsson and Hitt [11], in a paper written two decades ago, stressed that it took several decades for computers to prove their worth in business. What happens is that the new technology is introduced into a system largely run on more traditional methods, making it difficult for the novel technology to realise its full potential. In some areas, the established methods will remain because they are simpler, safer or cheaper, and big data applications will have to operate alongside them.

Big data was one of the many emerging technologies that were once on the so-called Gartner Hype Cycle [17]. The Gartner Hype Cycle considers emerging technologies to pass through five stages: Innovation Trigger, Peak of Inflated Expectations, Trough of Disillusionment, Slope of Enlightenment, and finally, Plateau of Productivity. In 2013, big data was listed as being in the Trough of Disillusionment phase, but since then it has been removed as an emerging technology; Gartner now considers big data to have firmly established itself.

Already, it is true, the use of big data has shown itself to be useful to a variety of businesses, for example, those in retailing, both online and conventional. But the use of IoT and big data in cities has hardly begun, and as shown in Chap. 2, although many cities label themselves as 'smart', it has so far made little measured difference in sustainability from similar (non-smart) cities. (Nevertheless, city officials of Barcelona have claimed that the use of IoT has enabled significant savings in water and lighting, and increased parking revenues [9].) It could well be that the 'Trough of Disillusionment' phase still awaits the specific application of big data in cities. As several researchers have stressed, city governance, with its many separate bodies for transport, environment, health and so on, is far more complex than the unified management of a corporation. The adoption and success of big data are likely to be uneven in the various sectors that are administered by urban governments.

4.7   Individual and Institutional Resistance to Big Data Solutions

The likely widespread future use of big data in some areas will often pass unnoticed, as it does today for machine translation or word completion online. In some science research fields, it is the only way of making use of the vast amount of measured data generated. In still other cases, though, it will threaten existing practices and professions, even if it does open up many new jobs in data analysis. For example, predictive policing is thought by some police officers to lead to deskilling of their jobs [53]. Even specialist medical professions are not exempt: radiologists and anatomical pathologists could in future see much of their work done by machine learning [41]. Are medical professionals prepared to give Artificial Intelligence control over diagnoses and perhaps even their patients' lives? At the very least, doctors want to see evidence that it is advantageous for patient health [49]. Moreover, we have already discussed the replacement of book recommendation editors by big data algorithms at Amazon.

The use of machine learning algorithms and big data could have an enormous impact on employment in another way—it could reduce opportunities for full-time employment. For several decades now, the end of work because of the IT revolution has been prophesied, but it has not happened. Some job categories, such as routine clerical jobs, in OECD countries at least, have largely disappeared, but have been replaced by other job categories. In many cases the new jobs created are part-time, with lower hourly pay rates and no job security. As Timothy Revell [47] has put it: 'The next big fight in the world of work may not be for shorter hours, but the right to a stable job'. AI in the workplace will also lead to deskilling, as already discussed, and less worker control over the job. Although overwork is stressful and unhealthy, regular work does provide benefits beyond the income earned. Even in situations where the application of machine learning could bring undoubted societal benefits, it will still be resisted if it lowers the quality of life for the affected workforce.

In other cases, big data will be simply seen as a misguided approach to the problem faced. We need to keep in mind a saying often attributed to Mark Twain: 'To a man with a hammer, everything looks like a nail.' Big data solutions, perhaps in the form of off-the-shelf computer packages, will sometimes be used inappropriately, and the resulting lack of success will slow its application in other problem areas where it can produce actionable results. In a widely read and cited article, Danah Boyd and Kate Crawford [10]—both of whom are incidentally affiliated with Microsoft Research—have posed some hard questions for big data. One risk they mentioned, not covered by the other research discussed in this chapter, is that big data could well change how we *define* knowledge, and how we even think about research questions. They also pointed out, as others have, that research always implies *interpreting* data, whether the data sets are small or large. The data do not speak for themselves.

Even when big data may be effective, its use in certain circumstances may be resisted. Susan Athey [5] has stressed that people often feel that it is important to know the reason *why* a decision affecting them was made, so that transparency may be more important than predictive power. In other cases, it might be important for emergency personnel like doctors 'to commit a decision rule to memory'.

An important question concerns the reaction of the public, who through their use of social media and credit cards, as well as energy, water, and transport systems will generate much of the data that will be analysed. Will they feel their data is just being used for the profit or benefit of others? Alternatively, will they come to decide that the application of this data is not only of personal benefit (for example, providing information on local traffic or public transport services, or detailed real-time pollution information) but is also necessary for cities to play their part in solving global environmental problems—particularly global climate change? Most likely, they will strike a balance, endorsing some uses and opposing others, as their familiarity with big data and how it is applied in each application grows. As always, media stories of abuses will likely force big data users to tread carefully, as has already happened with social media corporations.

4.8 Discussion

Big data could 'fundamentally change urban science' [20]. But can these varied problems facing the application of big data for the common good in cities be overcome? Both government and private industry have every incentive to make big data work. The remaining technical problems will most likely be solved if the past stunning successes of information technology are any guide. Martínez-Ballesté et al. [35] were also optimistic that both the privacy and security problems of big data in the context of smart cities could be solved. But as they stressed, if privacy concerns are not properly addressed, urban residents might avoid using smart city services. With the trust and support of the urban public lost, the many possible benefits from their application will not be realised. The same is likely to be true for commercial applications of big data. A code of ethics for privacy protection is needed, one that realises that as use of big data increases, there is a need to go beyond traditional privacy concerns, with its focus on data collection, and shift the emphasis to privacy concerns that can arise with using the data in new ways. And for all his penetrating critique of smart cities and big data, Rob Kitchin [28] has still argued that we need them—but that the ethical problems discussed above must be addressed.

Presently many national governments, while prosecuting private hackers, are themselves secretly involved in at least preparing for similar activities, as in cyber warfare. Widespread breaches of security, whether from private individuals, criminal organisations or governments put not only the use of big data, but also use of the Internet, at risk. Governments must continue to prosecute individual or criminal hacking, but must also realise that cyber warfare will be to no country's benefit. During the Cold War, both the US and the USSR realised that there could be no winners in an all-out nuclear war, and agreed to various forms of arms limitation. Something similar is needed for cyber warfare. If the potential security problems of big data are not solved, the use of big data for the urban common good will be severely restricted in its application. Bill Montgomery [37] has developed what he considered to be a solution to the security problem in an era of big data and the IoT. As he put it when describing this future world: 'The only way that the potential of IoT was fully realized was through the introduction of an IoT application-delivery model that established each individual nation as its own "trust center," while providing the ability for information to flow easily and securely from one trust center to another.' No data would be exchanged with non-trusted nations. It is still open to doubt whether this approach would fully solve security problems.

There are forces pushing for the introduction of big data solutions to urban problems. These forces include the actions of vendors with smart city software packages to sell, as well as the undoubted success of big data for several types of scientific research, and in marketing and retail. But as we have shown in this chapter, the greater use of big data also faces serious problems, which must be addressed if it is to contribute to urban sustainability. So far, its impact has been minor [21]. In the remainder of this book, we assume that the challenges to big data discussed in this chapter can and will be solved, that the new Information Technology, in general, does not prove to be too fragile for the present century [38]. We do not assume that the problems are entirely solved, which is probably impossible, but we do assume that with constant vigilance, they are reduced to manageable proportions. We similarly assume that (most) cities survive, and in future continue to account for a large share of the expanding global population.

References

1. Acquisti A, Brandimarte L, George Loewenstein G (2015) Privacy and human behavior in the age of information. Science 347(6221):509–514
2. Adlee S (2017) Your entire Internet history is now for sale. New Sci 234(3120):25
3. Allen N 2016. Cybersecurity weaknesses threaten to make smart cities more costly and dangerous than their analog predecessors. http://bit.ly/1Q33UTK
4. Anon (2017) Worldwide cyberattack. New Sci 234(3126):4

5. Athey S (2017) Beyond prediction: using big data for policy problems. Science 355:483–485
6. Anderson C (2008) The end of theory: will the data deluge makes the scientific method obsolete? Wired. June 23. http://www.wired.com/science/discoveries/magazine/16-07/pb_theory
7. Baraniuk C (2015) Forget carjacking, car hacking is the future. New Sci 227(3033):19
8. Bin Abdul Rahman MF (2016) Threats of driverless vehicles: leveraging new technologies for solutions. (RSIS Commentaries, No. 138). RSIS Commentaries. Nanyang Technological University, Singapore
9. Boulos MNK, Al-Shorbaji NM (2014) On the Internet of Things, smart cities and the WHO Healthy Cities. Int J Health Geogr 13(10):1–6
10. Boyd D, Crawford K (2012) Critical questions for big data: provocations for a cultural, technological, and scholarly phenomenon. Inf Commun Soc 15(5):662–679
11. Brynjolfsson E, Hitt L (1998) Beyond the productivity paradox. Commun ACM 41(8):49–55
12. Cartwright J (2016) Smartphone science. Nature 531:669–671
13. Cavoukian A, Jonas J 2012. Privacy by design in the age of big data. Accessed on 12 June 2017 at https://datatilsynet.no/globalassets/global/seminar_foredrag/innebygdpersonvern/privacy-by-design-and-big-data_ibmvedlegg1.pdf.
14. Crawford K, Calo R (2016) There is a blind spot in AI research. Nature 538:311–313
15. de Montjoye Y-A, Radaelli L, Singh VK et al (2015) Unique in the shopping mall: on the reidentifiability of credit card metadata. Science 347:536–539
16. Faghmous JH, Kumar V (2014) A big data guide to understanding climate change: the case for theory-guided data science. Big Data 2(3):155–163. https://doi.org/10.1089/big.2014.0026
17. Fisher S 2016. Gartner Hype Cycle 2016: blockchain, no big data. Accessed on 2 March 2017 at https://www.laserfiche.com/simplicity/gartner-hype-cycle-2016-blockchain-no-big-data/.
18. Fox P, Hendler J (2014) The science of data science. Big Data 2:68–70
19. Ganguly AR, Kodra EA, Agrawal A et al (2014) Toward enhanced understanding and projections of climate extremes using physics-guided data mining techniques. Nonlinear Processes Geophys 21:777–795
20. Glaeser EL, Kominers SD, Luca M et al (2016) Big data and big cities: the promises and limitations of improved measures of urban life. Econ Inq 56:114–137. https://doi.org/10.1111/ecin.12364
21. Glasmeier AK, Nebiolo M (2016) Thinking about smart cities: the travels of a policy idea that promises a great deal, but so far has delivered modest results. Sustainability 8:1122. https://doi.org/10.3390/su8111122
22. Gudivada VN, Baeza-Yates R, Raghavan VV (2015) Big data: promises and problems. Computer 48:20–23
23. Gumbus A, Grodzinsky F (2015) Era of Big Data: danger of descrimination. SIGCAS Comput Soc 45(3):118–125
24. Haight R, Haensch W, Friedman D (2016) Solar-powering the Internet of Things. Science 353:124–125
25. Hargittai E (2015) Is bigger always better? potential biases of big data derived from social network sites. Ann Am Acad Pol Soc Sci 659:63–76
26. Hvistendahl M (2016) Crime forecasters. Science 353:1484–1487
27. Jagadish HV, Gehrke J, Labrinidis A et al (2014) Big data and its technical challenges. Commun ACM 57(7):86–94
28. Kitchin R (2016) The ethics of smart cities and urban science. Philos Trans R Soc A 374:20160115. https://doi.org/10.1098/rsta.2016.0115
29. Kshetri N (2014) Big data's impact on privacy, security and consumer welfare. Telecommun Policy 38:1134–1145
30. Kshetri N (2016) Cybersecurity and development. Markets Global Dev Rev 1(2):3. https://doi.org/10.23860/MGDR-2016-01-02-03
31. Lazer D 2015. Issues of construct validity and reliability in massive, passive data collections. The Cities Papers.. Accessed on 7 March 2017 at http://citiespapers.ssrc.org/issues-of-construct-validity-and-reliability-in-massive-passive-data-collections/.

32. Lazer D, Kennedy R, King G et al (2014) The parable of Google Flu: traps in big data analysis. Science 343:1203–1205
33. Lee BG, Galli S, Brunner M et al (2011) New technical areas: exploring our future. IEEE Commun Mag 49(10):12
34. Mai J-E (2016) Big data privacy: the datafication of personal information. Inf Soc 32(3):192–199
35. Martínez-Ballesté A, Pérez-Martínez PA, Solanas A (2013) The pursuit of citizens' privacy: a privacy-aware smart city is possible. IEEE Commun Mag 51:136–141
36. Mayer-Schönberger V, Cukier K (2014) Big data. Mariner Books, Boston, New York, NY
37. Montgomery B (2015) Future shock: IoT benefits beyond traffic and lighting energy optimization. IEEE Consum Electron Mag 4:98–100
38. Moriarty P (1999) Don't depend on IT. Aust Quart 71(6):16–20
39. Mowshowitz A (2013) The end of the information frontier. AI Soc 28:7–14
40. Neff G (2013) Why big data won't cure us. Big Data 1(3):117–123
41. Obermeyer Z, Emanuel EJ (2016) Predicting the future — big data, machine learning, and clinical medicine. N Engl J Med 375(13):1216–1219
42. Oboler A, Welsh K, Cruz L (2012) The danger of big data: social media as computational social science. First Monday 17(7):60. https://doi.org/10.5210/fm.v17i7.3993
43. O'Grady M, O'Hare G (2012) How smart is your city? Science 335:1581–1582
44. Ornes S (2016) The Internet of Things and the explosion of interconnectivity. Proc Natl Acad Sci U S A 113(40):11059–11060
45. Perera C, Zaslavsky A, Christen P et al (2014) Sensing as a service model for smart cities supported by Internet of Things. Trans Emerg Telecommun Technol 25:81–93
46. Qualtrics 2010. The 1936 election – a polling catastrophe. 12 October. Accessed on 9 March 2017 at https://www.qualtrics.com/blog/the-1936-election-a-polling-catastrophe/.
47. Revell T (2017) Be careful what you wish for. New Sci 234(3125):22–23
48. Reynolds M (2017) Bias test to keep algorithms ethical. New Sci 234(3119):10
49. Rutkin A (2016) Medicine by machine. New Sci 231(3089):20–21
50. Salvatore C 2017. FDA reports hacking risk in St. Jude medical heart devices. Law 360. Accessed on 28 February 2017 at https://www.law360.com/articles/879525/fda-reports-hacking-risk-in-st-jude-medical-heart-devices.
51. Schmitt M, Marks P (2013) The right to bear cyber arms. New Sci 218(2912):26–27
52. Shah NH (2015) Using big data, Chapter 7. In: PRO P, Embi PJ (eds) Translational informatics: realizing the promise of knowledge-driven healthcare, health informatics. Springer, London. https://doi.org/10.1007/978-1-4471-4646-9_7
53. Shapiro A (2017) Reform predictive policing. Nature 541:460–462
54. Smith M, Szongott C, Henne B et al 2012. Big data privacy issues in public social media. 6th IEEE DEST Conference. Accessed on 27 March 2017 at http://ieeexplore.ieee.org.ezproxy.lib.monash.edu.au/stamp/stamp.jsp?arnumber=6227909.
55. Spratt S, Baker J 2015. Big data and international development: impacts, scenarios and policy options. Accessed on 25 April 2017 at https://opendocs.ids.ac.uk/opendocs/bitstream/handle/123456789/7198/ER163_BigDataandInternationalDevelopment.pdf?sequence=1.
56. Statistica 2017. Number of mobile phone users worldwide from 2013 to 2019 (in billions). Accessed on 16 March 2017 at https://www.statista.com/statistics/274774/forecast-of-mobile-phone-users-worldwide/.
57. Stutzman F, Gross R, Acquisti A (2012) Silent listeners: the evolution of privacy and disclosure on Facebook. J Priv Confid 4(2):7–41
58. Taleb N 2013. Beware the big errors of 'big data'. Wired blog. Accessed on 15 December 2015 at http://www.wired.com/2013/02/big-data-means-big-errors-people/
59. Tankard C (2012) Big data security. Netw Secur 2012:5–8
60. Utz S, Krämer NC (2015) The privacy paradox on social network sites revisited: the role of individual characteristics and group norms. Cyberpsychology 3(2):article 2
61. Wang W, Kennedy R, Lazer D et al (2016) Growing pains for global monitoring of societal events. Science 353:1502–1503

62. Waldrop MM (2016) More than Moore. Nature 530:144–147
63. Weinberg BD, Milne GR, Andonova YG et al (2015) Internet of things: convenience vs. privacy and secrecy. Bus Horiz 58:615–624
64. Wikipedia 2017 Closed-circuit television. Accessed on 24 February 2017 at https://en.wikipedia.org/wiki/Closed-circuit_television.
65. Wikipedia 2017. Criticism of Facebook. Accessed on 29 February 2017 at https://en.wikipedia.org/wiki/Criticism_of_Facebook.
66. Wong S (2015) How to take out a city. New Sci 227(3033):18–19
67. Ziegeldorf JH, Morchon OG, Wehrle K (2014) Privacy in the Internet of Things: threats and challenges. Secur Commun Netw 7:2728–2742. https://doi.org/10.1002/sec.795

Chapter 5
Big Data for Sustainable Urban Transport

5.1 Introduction

As we have discussed in the previous chapters, we are now living in a world where for the first time in human history, more than half of the world's population lives in an urban environment. In the next few decades (if present trends continue) perhaps 70% of people will live in cities [49], where people will require more mobility than ever. Big data empowered by cloud computing could make it possible for researchers and planners to deal with the growing challenges of urban transportation. The challenges we are facing are also one of the most important in this century: how can we enable urban dwellers in tomorrow's cities to travel more efficiently and sustainably, and provide services to the inhabitants, while maintaining comfort, pleasure, and safety during their journey? In this introductory section we will first give some detail on the types of big data potentially available in cities, then review the various options for sustainable transport.

5.1.1 The Portrait of a City from a Transport Viewpoint

As well as understanding personal activities, achieving real intelligence in a Personal Travel Assistant (PTA) needs a multi-disciplinary approach to gain an overall portrait of the city and its transport system. Also, data gathering should not be limited to a single source of information—the fusion of public heterogeneous data sources is becoming more important than ever. The data within the urban environment is all-encompassing, with data from different sources portraying the various aspects of the urban environment. For instance:

- The urban Points of Interest (POI) technology gives valuable information on urban functions, such as hospitals and residential areas

© Springer International Publishing AG, part of Springer Nature 2018
S. J. Wang and P. Moriarty, *Big Data for Urban Sustainability*,
https://doi.org/10.1007/978-3-319-73610-5_5

- Urban residents' activities in the social network not only present an urban crowd relationship map but also indirectly reflect the city's emotional and dynamic situation
- CCTV cameras collecting traffic image data etc. reflect the ongoing activities of the city.

Government and enterprises have realised the application potential of multi-source heterogeneous data fusion. For instance, the Shanghai government held the Shanghai Open Data Apps in 2015, which included the following data:

- Shanghai municipal Road Traffic Index
- Metro operation data
- Passenger card data
- Pudong bus real-time data
- air quality status data
- weather data
- road accident data
- elevated off-ramp data
- Sina microblogging traffic data.

The different types of urban data from all these multiple sources could help improve the accuracy of transport-related trajectory prediction. However, the multi-source heterogeneous data vary in attributes: for example, trajectory data is time and space data, monitoring and acquisition data are from images. How to manage and integrate large-scale multi-source heterogeneous data is the challenge of trajectory prediction. In the process of constructing a smart urban environment, the city's information infrastructure will be designed to provide much more and better information-enriched services. At the same time, this process accumulates a vast array of complex urban activity data.

Here, we briefly introduce the types of urban data that are commonly considered to be crucial for realising the goal of sustainable urban transport.

Where are you travelling now? As the premise for any location related discussions, until now GPS (Global Positioning System) provides the most crucial information for position recognition. A mobile device with a GPS receiver chip can collect information about the movement of objects in real-time such as the positions of people and cars in the city.

Is position the only information we can get from GPS? Currently there are several applications based on GPS in everyday use. For example, the most common one is Floating Car Data (FCD) technology to detect the traffic flow speed on the road. FCD uses the real-time data from the mobile phone in a vehicle being driven on the road network [51]. The mobile phones provide a data set which includes travel speed and direction as well as basic location data. FCD technology works as a sensing node for sampling the city's overall traffic situation.

The mobile phone is becoming an indispensable communication tool for daily life. It can provide many types of data: address book, call records, GPS positioning information, signalling records with the base station, internet records, and app use records. This data reflects the interest of residents in the city's activities, the activities themselves, the frequency of interactions, social relations, and other content—it has a vast application potential. A smartphone with a GPS receiver chip can also be used as a collection device for personal trajectories. However, due to privacy, security and many other issues (see Chap. 4) large-scale mobile GPS data collection is restricted, and current applications rely on volunteers for small-scale data collection and research. In the case study ISUNS project (see Sect. 5.6), the GPS data can be obtained from the mobile phone and navigation devices of participants.

Location Based Service (LBS) is an emerging network service mode in the mobile internet era. The data collected by LBS applications has clear geographical coordinates and combines the information features of traditional web services. It can be used for locating persons, navigation, and providing information on the nearest service, for example, an ATM. It thus can provide large amounts of information which can help planners more deeply understand the dynamics of managing a city.

One of the most important data types is maps integrated with POI data. Streets and buildings are the core framework of the city. The map is a basic way to describe urban structure, and the POI data is the core information of the city's functional units. Therefore, the city map and POI data is the smart city's most basic raw material, reflecting people's basic activities in the city.

Why are you travelling? The various means of transport the for daily commute all generate passenger data. Taxi passenger data can be obtained using the FCD GPS data with the passenger status of the taxi meter. Bus and subway passenger data can use the municipal traffic card credit card records. Passenger data contains much information on urban activities, which can be used in urban area function analysis, population flow detection, urban traffic system assessment, multi-vehicle human behavior research, urban traffic economics research and other fields.

Finally, Pan and colleagues [37] have reviewed how trace analysis and data mining from various sources can help improve both urban transport and urban planning. Data from mobile devices, vehicles, smart cards and floating sensors (RFID tags) can help gather detailed transport information to help both travellers (including urban visitors) and transport planners. Data on frequency of visits to various locations, for example, is not only useful for transport planning, but also to urban land use planners.

5.1.2 General Approaches to Sustainable Transport

Some of the embryonic existing uses for big data discussed in Chap. 3 could be rendered obsolete, and seen as merely stop-gap solutions for urban transport. Still, they could prove important during the transition to a more sustainable system. If, for example, large cuts in vehicle-km of travel were made in order to lower air pollution, oil use, and CO_2 emissions, the need for both traffic congestion management and parking spot information, as discussed in Chap. 3, would become largely irrelevant. However, a prerequisite for such a smart and smooth transport transition is a comprehensive understanding of the overall urban transport system. Hence this

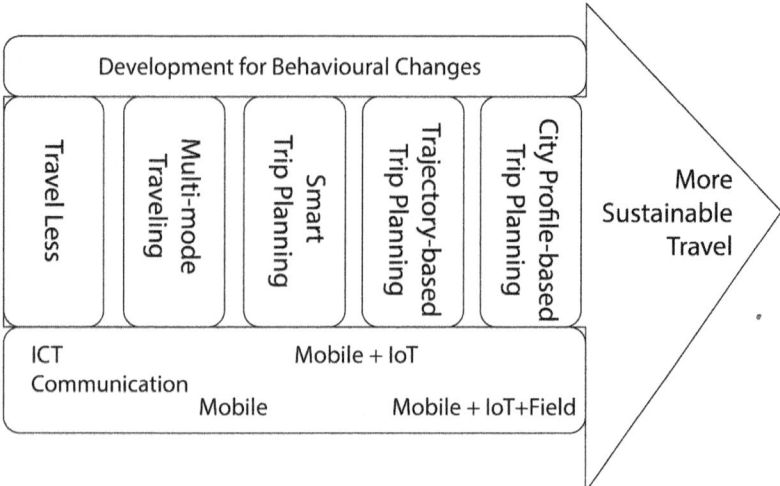

Fig. 5.1 Methods for more sustainable travel

chapter will mainly concentrate on the other aspects of smart transport, examining how big data can reduce car travel demand itself by either encouraging a shift travel to other modes, or by reducing the demand for vehicular travel overall.

In this chapter, we will look at how big data could be applied far more widely, with far greater effect, to one of our main themes, urban transport. Figure 5.1 summarises the general approaches possible for both reducing the environmental impact of urban travel, while at the same time enhancing the welfare of urban travellers:

- Travelling less, enabled by ICT communications
- Multi-mode travelling, enabled by navigation systems
- Smart trip planning, enabled by mobile devices with real time update
- Trajectory-based [38, 41] trip planning, enabled by combination of mobile and IoT technologies
- City profile-based trip planning, enabled by a comprehensive usage of big data from other sectors as well as transport, and covering the entire city.

As a first task, how can big data help cut the energy consumption of travel in cities? Funk [13] provided an insight into why big data (or IT generally) can help reduce both the carbon emissions and resources used in transport. He pointed out that the annual present gains in energy efficiency are far lower than the efficiency gains experienced in IT (as enshrined in 'Moore's Law'), suggesting that more use of IT in transport will pay large dividends in sustainability.

Chapter 1 introduced the several means for reducing urban travel energy (and consequently GHG, air, and noise emissions), which are repeated here:

- by shifting to more energy efficient modes, namely public and non-motorised transport

- by reducing the demand for urban transport, using IT to either combine previously separate trips, or to substitute for the trip itself
- by increasing the occupancy rates on both public and private vehicular transport
- by increasing the vehicle efficiencies (vehicle-km per primary MJ) for both public and private vehicular transport.

The rest of this chapter will build on the discussion in Sect. 1.3.2, and will examine the potential role of big data in each of these approaches (Sects. 5.2–5.4). Since the potential for big data is much greater for the first two approaches, they are discussed in greater detail. In Sect. 5.5 we discuss traveller *comfort* on public transport, which is a significant problem for all high-density cities in Asia, and in low-income cities generally. Section 5.6 describes a personal travel assistant developed by the authors and their colleagues [49] for use in Beijing. As we have already stressed, most of big data's potential lies in the future. The concluding section (Sect. 5.7) accordingly looks at future urban transport and summarises the potential role of big data in meeting sustainability goals for transport.

5.2 Shifting to More Energy Efficient Modes

Urban public transport modes are usually more energy efficient than car travel [35]. Reasons include higher occupancy rates, particularly at peak travel times, lower rolling friction for steel rail transport, and the greater values of seat-km per MJ of primary energy that larger vehicles enjoy. Electric public transport also has the potential for regenerative braking—as of course do hybrid and battery electric vehicles. Non-motorised transport is even more energy efficient as well as being superior to vehicular transport in reducing land use, air and noise pollution, traffic casualties, and GHG emissions [26]. It is also far cheaper, and available without the need for a driver's licence. At present, transport priorities are car first, then public transport, and finally non-motorised travel—if the latter is recognised at all. For sustainable and healthy urban futures, we need to reverse this order and give priority to non-motorised transport.

5.2.1 Public Transport

A number of cities have already implemented smart cards for automatically debiting fares on their public transport systems, such as the Myki card in Melbourne, Australia, and the Oyster card in London. They are well-accepted by users. Michael Batty [3] has discussed the promise and difficulties of using data from the Oyster card system for transport analysis. Over 6 months in 2011–2012, researchers at University College, London collected around one billion records of travellers who 'touched' on and off. Unfortunately, despite the enormous amount of data available, only about 85% of the customers of London public transport use the Oyster card, which raised questions of how representative even this huge sample was. Of these 85%, 10% did not touch off due to open barriers. He pointed out that several assumptions must, therefore, be made to make the data useful to transport planners. Nevertheless, such data has great potential in helping planners understand travel patterns for regular users of London's public transport (the 85%), and, as more data accumulates over time, to see what changes in travel patterns are occurring.

For Beijing City, Long [53] suggested that it is possible to analyze the bus cardholder's place of residence, employment and commute travel characteristics in Beijing, based on a week of credit card records combined with the 2005 Beijing residents travel survey and land use map data from Beijing's 8.5 million bus cardholders in 2008.

In Brisbane, Australia, Tao et al. [45] have used smart travel cards to evaluate the spatial-temporal dynamics of Bus Rapid Transit (BRT) trips in comparison with trip making on the non-BRT bus services. Travel data were collected for various days representing different travel patterns: a workday, a Saturday, a Sunday, a school holiday, and a public holiday. As with the smart card systems already discussed, the Brisbane card provides both user details, as well as specific trip details such as the boarding stop, time of touch on of the smart card, and bus route number. Marked differences were found between BRT and ordinary bus travel. The authors concluded that: 'The results offered more detailed and reliable insights into BRT use than had been previously attained. From this, important implications for evidence-based BRT policies were identified to inform service management (e.g., service monitoring on critical pathways, establishing flexible route allocation mechanism) and infrastructure provision (e.g., enhancing the utility of current busway, selecting potential BRT corridors).'

Two further examples come from Boston, USA. In 2015, city officials started receiving quarterly reports from Uber, the car booking company, on the origin and destination of all Uber trips, the time of day and date when they were made, and the distance travelled [18]. It was hoped this data would help planners better understand the transport flows in that city. This knowledge would also help public transport network planning. In a second Boston study based on the effects of widespread ride-sharing, Alexander and González [2] tried to answer the question: Does ride-sharing increase or decrease urban traffic congestion? They showed that ride-sharing would generally decrease congestion—in effect, that ride-sharing would not be at the expense of public transport or non-motorised trips.

One possible reason why public transport patronage is low in many OECD cities is that many travellers, especially regular motorists, are unfamiliar with both the services available and the timetables. Also, at present urban travellers largely rely on *travel routines* to make decisions about which route to choose and which mode to use [17]. Particularly for car travellers, the existing routines will have to change if transport is to become more sustainable. Smart phone apps are increasingly making real-time data for all modes available to travellers and informing them of any delays [26]. In Melbourne, for example, the available public transport app gives the times of the next five arrivals at the public transport stop at which the traveller is waiting.

Miller [25] has outlined an ambitious plan for the use of big data for urban passenger transport. He pointed out that, at present, particularly in the US, there is a *transport monoculture*: it is assumed that all trip types, to all destinations, at all times, are best done by car. Instead, he urged a *transport polyculture*, in which the various forms of non-motorised transport, and public transport, are seen as vital parts of the transport mix. His idea was that the various transport modes should cooperate to achieve environmental and other aims, rather than be in conflict, as at present. Innovations such as Seoul's Personal Travel Assistant (PTA) are a step in this direction, presenting all modal choices for each trip undertaken. This journey planner allows travellers to select for any trip the travel mode which will variously give the fastest trip, the cheapest trip, or the trip with the lowest GHG emissions [4].

5.2.2 Non-motorised Transport

Non-motorised transport—walking and cycling—will have to play a far larger role in urban transport if transport GHGs are to be significantly reduced. In an urban context, these modes have great potential for expansion, given the significant share of urban trips which are less than 2 km. In the UK, at least, the share of trips by non-motorised transport has fallen in recent decades [12]. Increasing the use of these modes can help cut urban transport air pollution, noise, GHG emissions, and energy use. Their increased use can also reduce the 'serve passenger' trips associated with chauffeuring children to school or activities by car.

The existing use of non-motorised travel varies greatly between the world's cities, with the greatest use being in African and Asian cities, and the lowest in those of the US; it is accordingly often stigmatised as the poor person's mode of choice. Still it can also be high in wealthier cities, too, suggesting that there is potential for improvement. For the year 1995, Cameron et al. [8] estimated that 34% of all trips in Hong Kong were by walk/cycle, compared with only 5% in Phoenix in the US. The share of non-motorised travel in the lower-income cities of Asia would be similar to, or greater than, the share in Hong Kong. In the cities of Africa and much of Asia, where national car ownership rates can be as low as 10–20 per 1000 residents (it was only 15 per 1000 in India in 2015 [36]), most trips are probably still by walking and cycling.

The other major benefit from far more use of non-motorised modes is for urban health, a topic dealt with in more detail in Chap. 7. Exercise has been termed the 'wonder drug' by Andy Coughlan [10], because of its variety of health benefits. He wrote: 'A plethora of recent studies shows that exercise protects us from heart attacks, strokes, diabetes, obesity, cancer, Alzheimer's disease and depression. It even boosts memory. And it has the potential to prevent more premature deaths than any other single treatment, with none of the side effects of actual medication.' Given its benefits in so many areas, it is surprising that it still has had so little real support from transport or health policy makers.

How could big data encourage more non-motorised travel? One way could be through a PTA. For example, travellers could be advised how long a given trip could take by walking or cycling, based on the traveller's average walking/cycling speed as calculated from personal smart phones. The PTA could also give the estimated personal energy expenditure for the proposed trip, useful if the traveller had a daily exercise budget, and for the environmentally conscious, the CO_2 saved by forgoing the car option. It could also assess whether the estimated trip time would fit in with the time available, such as the lunch hour, or before an appointment, etc. Further, based on data from street and pavement sensors, at traffic lights, the green time for pedestrians could be extended if the volumes of pedestrian traffic warranted it.

5.3   Reducing the Demand for Urban Transport

Big data-supported PTAs could support modal shifts, which would help ease the urban transport sustainability challenge to some degree. However, large cuts in urban transport energy use will most probably require corresponding cuts in vehicular mobility itself—not an easy task [27, 32]. One reason why reducing vehicular transport is so difficult is that its present cost structure in most cities is not conducive to energy conservation. In a paper comparing Tokyo and Shanghai, Luo et al. [21] showed the higher parking restrictions and costs resulted in far fewer daily trips per car in Tokyo. Perceived costs of car travel, mainly fuel costs, are a small part of the total costs of owning and operating a car, since purchase, annual insurance, and registration costs are independent of vehicle kilometres. For urban freight vehicles, especially smaller ones, driver costs make fuel costs an even smaller fraction of total costs. With data collected on distance travelled by class of road and time of day for each vehicle, it would be possible to move to a 'pay as you drive' scheme for both passenger and freight road vehicles, which would make vehicle owners much more aware of the real costs of road transport. Presently unpaid environmental costs of vehicles—their negative externalities—could also be included, for example, by a carbon tax based on fuel consumption.

In imposing such a scheme, it would be important to avoid increasing inequality in urban areas where car travel is dominant, since in many OECD cities at least, low-income households often live in outer areas, where they not only have greater travel needs, but also have lower quality public transport services compared with inner urban households [27]. (In contrast, in the low-income, low car ownership cities of many industrialising countries, raising the costs of private motoring could enhance equity.) Nevertheless, the huge subsidies fossil fuel energy presently receives [30], including those for transport, will eventually have to be greatly reduced. Already, OECD countries in Europe and particularly Asia which have low domestic fossil fuel reserves, such as Sweden, Finland, Japan and South Korea, tend to have higher transport fuel costs and less fuel consumption than the US or

Australia, with their vast fossil fuel reserves [5, 29]. Concerns about supply security, together with lower fuel use, seems to make higher prices more acceptable.

Car insurance and registration rates could be based on not only vehicle-km but also on time and location of travel (with higher rates for travel in congested inner city zones, as is already the case for the London and other road pricing schemes [50]). But with the detailed data on driving speeds and acceleration increasingly available from vehicles, it would even be possible to incorporate driving practices into car insurance rates.

5.3.1   Reducing Travel with IT

Over the last decade or so, many OECD cities have seen a fall in average vehicular passenger-km per capita [27]. At the same time, there has been an explosive growth in both smart phone ownership and social networking. Some researchers (e.g. Lyons [22]) have therefore considered whether the rise in new forms of IT have at least partly *caused* this documented decline in personal urban travel. The question arises: could big data facilitate urban travel reductions through IT *substituting* for travel? This idea goes back at least four decades, with a book-length publication on tele-work (also called tele-commuting) in the US by Jack Nilles [35]. The aim of tele-working in the 1970s, with its oil crises, was to reduce both traffic congestion and transport energy consumption. Subsequently, the concept was extended to tele-medicine, tele-education (and cyber universities), tele-shopping and so on. Although today, IT has been applied in all these areas, it is fair to say that its impact on urban travel so far has been negligible. Historically, of course, telecommunications and vehicular travel have both grown in step, prompting some researchers to argue that the two are complementary rather than competitors [9, 31].

One view is that although use of the latest ICT, specifically mobile phones, may reduce travel in some cases, it could increase travel in other cases [1, 12]. Dal Fiore et al. [11] argued that 'mobile technology might offer people numerous new reasons to be mobile: by making them more informed; more capable of using a larger variety of physical spaces and re-negotiating obligations in real-time; and potentially more efficient in the allocation of their travel time and resources.' However, they also added that the technology could in some cases work to make travel less appealing.

What we also have to take into account is that ICT itself will change the travel patterns it is supposed to replace. Think of how ICT has the potential to change the nature and location of work/employment, health, education, shopping, entertainment. So far, we have only considered ICT effects on *existing* trip patterns, destinations, etc. Hence the past, minor, impact of ICT on urban travel volumes, patterns, and modes may be a poor guide to the future.

As already mentioned, we consider the application of big data to urban sustainability as a necessary but not sufficient condition: policy support is also needed. For large urban car travel reductions, it will also be necessary to increase both the monetary costs of road travel by application of carbon taxes (discussed in more detail in

Sect. 6.1), and increased tolls and parking charges, and to reduce the *convenience* of car travel [27]. The convenience of using the car in urban areas can be lowered by measures such as reducing speed limits and inner urban parking spaces, ending arterial road building, and road closures in the central business district (CBD). The main effect of these latter measures will be to raise travel times for trips by car. Raising both monetary and time costs can be justified by the accompanying reduction in collisions, traffic noise, community severance, air pollution, oil dependency and GHG emissions. Although all these measures could be quickly implemented, for equity reasons raising the travel time costs should be implemented first, as all travellers have 24 h per day, but incomes are most unevenly distributed, and as discussed above, lower income households often have greater travel needs in OECD cities.

Importantly, big data facilitates analysis of many *natural experiments* in urban transport. The effects of changes in road speed limits, parking restrictions, road closures, road charges—all regardless of whether temporary or permanent—can be analysed, as well as the effects of introducing new bus services, bus-only lanes, and high vehicle occupancy lanes. Unlike domestic energy use, per capita transport use (even after adjustment for income and other household characteristics) also varies spatially across the city, with generally higher levels of per capita car travel in outer areas of OECD cities, and lower public transport use. It also varies by income level, by age group, and by employment status. Trying to incorporate all these variables into conventional sample surveys is difficult because obtaining reliable data for subgroups (for example, low-income elderly householders in the outer suburbs) requires very large (and thus very expensive) sample sizes.

Nevertheless, some present applications of smart transport can be counterproductive from an ecological sustainability viewpoint, in that they can lead to increased urban transport energy use and GHG emissions. This counter-intuitive result arises from feedback effects. Very congested cities such as Tokyo and Hong Kong have low average travel speeds for road traffic, with much stop-start driving. But, on the other hand, they have very well-patronised rail systems, and far greater use of non-motorised transport than in North American or Australian cities (see Sect. 5.2.2). High levels of rail patronage (and high rail seat occupancy rates) result in public transport being several times as energy efficient as car travel on a passenger-km per primary energy basis [34, 35]. Overall, car energy use and GHG emissions *per vehicle-km* might be higher than in lower density cities but are more than compensated by far lower car vehicle-km per resident.

The following is a speculative view of how big data could help reduce vehicular travel in the presently car-dominated cities of the OECD. Travel reductions can occur in several possible ways: some trips can be foregone; some can be combined with another (non-discretionary) trip; and some can be made to closer destinations. Only the last two are of interest here. For combining two or more previously separate trips, the intending traveller could input the required information into their PTA app, such as the start and finish times for work and its location, or the time and location for picking up children after school, and any other constraints. These constraints could be designated as constant for every weekday. For trips that are discretionary for both their timing and location (e.g. shopping trips), the user could input various acceptable options. The PTA could then advise the traveller on ways of minimising total travel time, money cost, or CO_2 emissions for that day.

A suitably designed PTA could also help with shortening individual trips. Trips can be divided into two types: discretionary and non-discretionary. Discretionary trips into trips like shopping, recreation and social visiting for which time and/or location can be varied. Non-discretionary trips include work and education trips, which, in the short term at least, must be made at a definite time to a definite place. Especially in the cities of North America and Australasia, the rise of the car has brought about a suburbanisation of trip destinations such as shopping, entertainment, and workplaces. Paradoxically, this increased opportunity for localising activities was accompanied by a steady rise of per capita vehicular travel in most cities [27].

Evidently, the perceived low time and money costs of car travel made such extra travel acceptable. But as the need for travel reductions is increasingly recognised by the urban public, and supported by policies such as carbon pricing, travellers will increasingly prefer *local* destinations—for shopping and social activities, and over time for work as well. As shown, the PTA will be valuable in the perhaps long transition period, but after individuals (or households) have established a new travel routine, it will be less necessary for day to day travel.

5.3.2 Reducing Freight Transport

Reductions in kilometres travelled are also needed for freight transport vehicles. Often this is possible by eliminating 'wasted travel', the unnecessary km of travel caused by sub-optimal routes. This is especially a problem when a freight vehicle has to make multiple deliveries on the one trip, with locations that vary each day. Victor Mayer-Schönberger and Kenneth Cukier [23] have described how in 2011 the US freight company UPS was able to cut the travel of its vehicles by about 48 million km, with attendant fuel savings of over 11 million litres. The reduced km of travel and fuel translated into lower costs and fewer accidents. One reason for these savings was that the algorithm used 'compiles routes with fewer turns that must cross traffic at intersections', which in turn reduces the time taken, fuel use, and accidents.

On-line shopping is often viewed as a way of cutting vehicular travel to shops. One problem is that even if passenger vehicle-km is thereby reduced, the reduction will be offset by a rise in freight transport deliveries. What is more, these deliveries will usually be in low load capacity trucks, which have low energy efficiency per tonne-km of freight carried. Can IT help solve this problem it has created?

Taniguchi et al. [44] have argued that there is a variety of ways in which the new technology can help reduce the environmental costs of freight travel. They pointed out the limitations of existing road pricing schemes: cordon-based pricing schemes encourage speeding to minimise time in the cordon. There is no one objective for freight vehicle pricing, which could include cost recovery for infrastructure, reducing peak hour travel by trucks, and ensuring high vehicle loading rates. However, while road pricing schemes might be the best way forward in the transition period to more sustainable cities, in the longer term reducing energy, GHGs and air pollution from freight will become most important. Charges would then be based on the environmental costs of freight, with the levy perhaps added to fuel costs.

One suggestion for achieving reductions in the environmental burden of the urban freight system is 'co-modality'—using passenger public transport vehicles at off-peak times for freight delivery [44]. The present distribution of warehouses is presumably optimal for distributors but would change if the external environmental damages had to be internalised. Environmental taxes for freight would, *ceteris paribus*, raise freight costs, but if these and other initiatives just discussed were also introduced, overall freight costs in a given city might even fall because of reduced freight vehicle-km and reduced driver costs.

5.4 Raising the Energy and GHG Efficiencies of Urban Transport

Measures to improve energy and/or GHG *efficiency* in the transport or other urban energy use sectors can only take us so far and are in any case subject to the *rebound* effect [28]. Rebound in energy use occurs because improving the energy efficiency of, for example, cars, lowers the per kilometre costs of motoring, leading to more kilometres being travelled. In cases where demand is inelastic, indirect rebound can still occur, because the money saved from higher energy efficiency and resulting lower fuel costs can now be diverted to the purchase of other energy-using goods or services. As stressed in Chap. 1, the urgency of climate change will probably require reductions in the *use* of energy-using devices, as well as efficiency improvements. Raising the price of transport energy (by either changing the structure of motoring costs or imposing a carbon tax, as discussed in Chap. 6) can also prevent energy rebound. In this section, we first consider the potential for automated vehicles, then look at approaches to raising the energy efficiency of vehicles by reducing wasted travel, congestion, etc.

5.4.1 Automated Vehicles: A Possible Solution for Improving Efficiency?

The idea of fully automated, driverless vehicles has a long history: General Motors trialled one such vehicle in 1935 [39]. In the 1990s, vehicle automation took the form of Intelligent Highway Vehicle Systems (IHVS). In IHVS, groups of instrumented vehicles would move in close formation on instrumented freeways. Trials were conducted in the US, as well as in Europe and Japan. Proponents claimed that IHVS would both save both fuel and increase road capacity because the close spacing would reduce air friction on all following vehicles in the 'platoon' and reduce road space per vehicle. On non-instrumented roads, vehicles would need to revert to manual control.

Today, fully automated vehicles (AVs), such as the Google car, have been licensed to travel on roads in some US states, and millions of vehicle-km of travel have been logged. The justifications for widespread AV introduction include those for IHVS, but some new advantages are claimed. Fuel efficiency would be enhanced not only because of closer spacing on all roads (not only freeways) but because vehicles could now be redesigned. In the best case of vehicles never having to revert to manual control, the driver-controlled steering and braking systems could be dispensed with, as well as the need for forward-facing front seats. Advocates also claim that AVs would be far safer than driver-controlled vehicles, because in 90% of today's collisions, the driver is at least partly in error [48]. With this higher safety margin, even further weight reductions would be possible, because the vehicles will not need to be as crash-resistant as present vehicles. AVs would thus be both safer and more energy and GHG efficient.

Lawrence Burns [6] has stressed another possible advantage of AVs—shared car ownership. Although shared ownership has always been possible, driverless vehicles could come to be regarded as taxis or hire vehicles. Privately owned cars are presently parked 90% of the time, allowing up to an 80% reduction in vehicle numbers, thus greatly reducing both the parking space needed and the embodied energy costs of car manufacturing. He also envisioned AVs as being electric drive, but, of course, this is possible without AVs.

But would AVs deliver the above-mentioned energy savings, or will AVs instead greatly *increase* vehicle-km of road travel, swamping any possible savings from more efficient vehicles? For a start, fully-automatic driving on all roads under all weather conditions will not happen any time soon—perhaps not even before 2030 [16]. Until then even AV-capable cars will need to be designed to enable functioning in both driver-controlled and automatic modes, making them heavier, and so less fuel efficient, than existing vehicles.

Even if the world switches completely to AVs and they do prove safer than existing vehicles, energy reductions are not assured. We have already seen that, the perceived costs of travel include both a time and a money component. But what if AVs allow both to be cut? Nikolas Thomopoulos and Moshe Givoni [46] have presented data for the US showing that total time spent driving, both passenger and freight vehicles, in the US amounts to some 75 billion hours per year. Driverless vehicles could thus generate annual productivity gains which they estimated at $507 billion. Cost savings would also result from fuel savings, as already discussed. The resulting perceived drop in both the monetary and perceived time costs of travel could greatly *increase* private vehicular travel and undercut any existing advantages of public or non-motorised urban transport.

Further, car ownership and sales could rise instead of fall, for two reasons. First, there would be no need for a driver's licence, so even young children—with wealthy parents—could now have a car. Second, because car ownership and operation costs would fall, insurance premiums, fuel costs, and car purchase costs should all be much lower. Nevertheless, given the likely safety problems AVs will face—particularly for

Asian cities were much urban travel is on foot, or by pedal cycle, electric cycle or motorcycle, modes that cannot be automated, it seems unlikely that AVs can contribute to the sustainability of future cities. As Eric Bruun and Moshe Givoni [7] have stressed, getting people to give up driverless vehicles may be even harder than getting them out of conventional cars. Above all, AVs may never happen, as there are so many social, legal and technical problems still to be resolved [15].

5.4.2 Improving Vehicle Efficiency

As seen in Sect. 5.4.1, some have thought that switching to AVs, by enabling a radical redesign of the vehicle, would lead to a much lighter vehicle and thus less fuel use. However, if the driving task is dispensed with, seats could face each other, and cars could now be seen as mobile offices or entertainment centres. In such cases, vehicles could become larger, with even less fuel economy than conventional vehicles. Here we see that not all proposed uses of big data will automatically further the aims of urban sustainability. In some cases, at least, we will need to turn our backs on the possibilities opened up by big data in the interests of sustainability.

Fuel efficiency in vehicles depends not only on the physical characteristics of the vehicle, such as weight and engine efficiency, but also on how they are driven, and the traffic conditions under which they operate. Big data can help the motorist choose a route, based on current traffic conditions, that will minimise fuel loss caused by congestion, and wasted travel caused by non-optimal routes. It can also advise the motorist on driving in an 'eco-friendly' manner. Such fuel savings, while useful, will only be of temporary benefit if we must transition in future to a passenger transport system based largely on public and non-motorised transport.

Nevertheless, we should not expect too much from energy efficiency gains. Historically, there have been continuing efficiency gains for each vehicular mode, but at the same time a progressive shift to faster and more energy intensive modes. Thus, at the global level, public transport largely replaced non-motorised travel, car travel replaced most public transport, and now air travel has made inroads into long-distance car travel.

5.5 Beyond Energy Efficiency: Traveller Well-Being and Comfort

Reducing the environmental, resource and health-related negative impacts of urban transport is vital, but is not the full story. Urban travellers must also be able to make their journeys in comfort and safety, and in a reasonably timely manner. It is also desirable that public transport services be relatively frequent to avoid long waiting times, and that interchanges from one public transport service to another if needed, are convenient and fast. On a passenger-km basis, public transport modes are in general superior to car travel for both primary energy consumption and GHG (and air pollution) emissions. The energy efficiency of public transport varies greatly from city to city, but in general is far higher in Asian cities—regardless of their average per capita income level—than in the cities of North America, Australasia, or even Europe. The main reason is the far higher urban density of Asian cities, and their resulting lower levels of both car ownership and car share of urban travel, both of which drive up patronage—and occupancy rates—on public transport services.

In the cities of Australasia and the US, urban public transport seldom carries more than 10% of all vehicular travel. In contrast, even in the high-income Asian cities such as Hong Kong and Tokyo, public transport can account for over half of vehicular passenger-km [34]. Very often, the public transport system is already over-loaded—in Tokyo, the number of total annual vehicular trips, whether by public transport or car, has not risen for decades [43]. Unlike US/Australasian cities, the present problem is not encouraging more travel by public transport, but getting more value from the existing system. In the longer term, all cities—at least those in high and middle-income countries—will need to reduce overall vehicular passenger-km. However, even in the cities of China and India, there is room for further expansion of public transport systems if they can help reduce burgeoning car travel.

In OECD countries at least, including high-income Japan, average car seat occupancy rates are around 30%, or 1.5 persons for a five-seater car, with occupancy at peak hours being lower than at off-peak times. In marked contrast, Asian public transport services—and even peak hour services on public transport in non-Asian OECD cities—very often have seat occupancy rates far exceeding 100%. In other, words many passengers are standing, often at crush volumes. This crowding results in very high energy efficiencies, but at the expense of passenger comfort, particularly in hot climates. 'In Beijing in 1995, public transport energy efficiency (passenger-km per MJ of primary energy) averaged over all modes, 'was about six times that of private car travel' [49].

This overcrowding is exacerbated by prolonged travel times—for all modes, including non-motorised ones. Bangladesh, for example, has one of the lowest car ownership levels in the world. Nevertheless, its megacity capital, Dhaka, is one of the most traffic congested cities on the planet (see Fig. 5.2). Even in this low-income city, information technology is being implemented to assist traffic flow and control, and smart cards have been introduced on some bus services [19].

5.6   Case Study of a Personal Travel Assistant for Beijing[1]

5.6.1   Background

In Sect. 5.2.1 we discussed a personal travel assistant (PTA), in the context of increasing public transport ridership by providing relevant travel time, energy use and carbon emissions for the various competing modes. On their own, such

Fig. 5.2 Traffic congestion in Dhaka, Bangladesh. ['World Class Traffic Jam' by b k available at http://bit.ly/2uOhpU1 under a Creative Commons licence Attribution-ShareAlike 2.0 Generic (CC BY-SA 2.0). Full terms at http://creativecommons.org/licenses/by/2.0]

approaches are unlikely to give the reductions needed. For example, a small pilot study in Switzerland showed that modest increases (14%) in 'sustainable transport choices' were possible with a mobile phone app that test subjects used over one month [14]. However, the authors of the study warned that longer interventions would be needed for more significant increases in the use of sustainable transport.

In this section, we describe a PTA developed for Beijing, but with a difference. The aim is to reduce energy (and hence CO_2 emissions) from Beijing passenger travel by using the idea of a transport energy quota for each resident. An analogous, but more inclusive proposal was considered for the UK: The Personal Carbon Trading scheme. As described by Richard Starkey [42], this proposal would 'allocate equal tradeable carbon credits to all eligible individuals (who may simply be taken here as all adults), with the aim of reducing carbon emissions, as part of an emissions trading scheme'. Our proposal would be restricted to one city and passenger transport, but would likewise allow each 'eligible person' (however defined) a monthly transport energy quota in MJ. Although we use energy for the quota, the scheme could just as easily use kg of CO_2 equivalent. China already has in place several regional pilot emission trading schemes which could eventually form the basis of a nation-wide scheme [20]. The idea of a personal transport carbon allowance should be a good fit to national carbon pricing.

In our paper on a PTA for Beijing [49], we gave details of the algorithm used to compute energy costs for each trip. Briefly, the algorithm calculates the energy cost of each trip, taking into account the mode used, and for car travel, the traffic conditions likely to be encountered for that route at the time of travel. Non-motorised modes are assigned zero energy cost. At any time during the month, the user can calculate the remaining energy quota, and vary the travel mode for the remaining days to stay within the quota.

In Sect. 5.6.2 we describe the application of such an energy credit system through a series of simulations. The ultimate aim would be to implement this system on mobile phone based navigation software which is supported by the Urban Transport Energy Saver (UTES) database on the Azure platform. We also give a rough estimate of the data needs of such a system, assuming a modest level of use.

5.6.2   Description of the Application[1]

The UTES tool suite discussed here aims for general usability by providing a broad set of interface options, but with a particular focus on pre-trip information as opposed to in-trip guidance. The ultimate goal for UTES is to develop a platform whereby every urban sensor, device, person, vehicle, building, and street can be potentially used to probe city dynamics to enable city-wide computing that serves the population's travel activities. UTES aims to enhance urban transportation through an iterative process of sensing, data mining, understanding and improving urban transport systems. In the following sections, the system architecture, design, and implementation of UTES are discussed. The design of UTES focuses on interactive visual analytics to provide the user with the most appropriate information. The design will eventually consider factors such as urban residential density, travel styles, alternative options, costs, time restraints, travel experience, etc.

In contrast to traditional urban transportation systems, UTES would leverage the *cloud* to profile and optimize data classifiers for mobile devices, depending on the current device context and sensor data characteristics, to provide interactive visual analytics for supporting decision making. This project focuses on analysing and mining dynamic information.

Recently, there has been a trend to using a combination of real-time Global Positioning System (GPS) and Radio Frequency Identification (RFID) sensors deployed in vehicles, together with static city sensors and cloud service to optimize eco-efficiency. This approach can benefit from the 'city portrait' concept by embracing a wide range information from static sensors such as street cameras, traffic lights location and cycles, speed of cars, road lighting, temperature and general weather information and vehicles themselves, including car speed and GPS location information. This allows optimal management of vehicles and traffic in real-time, potentially improving system energy efficiency.

The UTES application will be developed and examined through a series simulations based on real Graphical Information System (GIS) data, potentially with real-time data feeds. As an example, Figs. 5.3 and 5.4 show an indicative travel simulation

---

[1]This section is a revised and updated version of Sect. 2.2 in Ref. [51]. Used by permission of Elsevier.

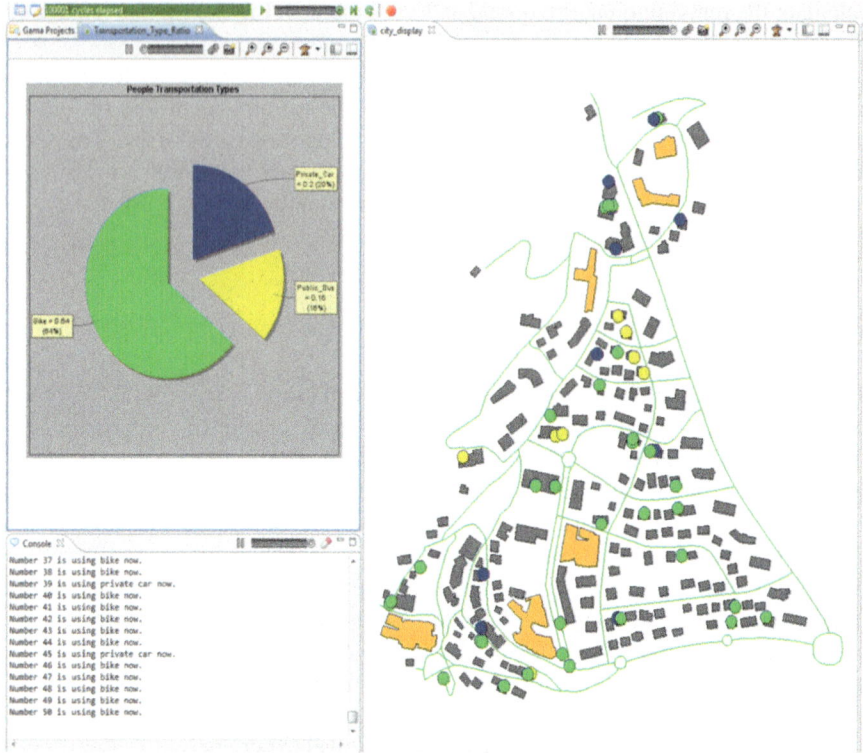

Fig. 5.3 UTES project, screen display for travel simulation for a given trip (color figure online)

within a simulated urban environment for a given trip. In the window at the right of
Fig. 5.3, the following elements are presented:

- A GIS map, with the lines presenting roads in the area. The color of the roads
  changes with traffic volumes, with green representing lightly-trafficked roads.
- Buildings are currently classified on the map as either public/business (yellow)
  residential (grey)
- Users (current travellers and trip planners) are represented by circular dots on the
  map; there are 50 users in total in this simulation
- Transport modes are represented by different circle colours This simulation has
  three modes: private vehicles (blue), public transport, i.e. buses, trains or underground rail (yellow), bicycle/walk (green).

The top left graph in Fig. 5.3 gives the transport mode share: blue for car (20%),
yellow for public transport (16%), green for walk/bicycle (64%). The bottom left
corner lists the mode currently used for each traveller. Figure 5.4 gives the energy
consumption of the overall transportation system by those presently travelling. The
blue curve shows maximum energy consumption assuming all travel is by private
vehicles, the red curve energy consumption optimized by UTES method. As the

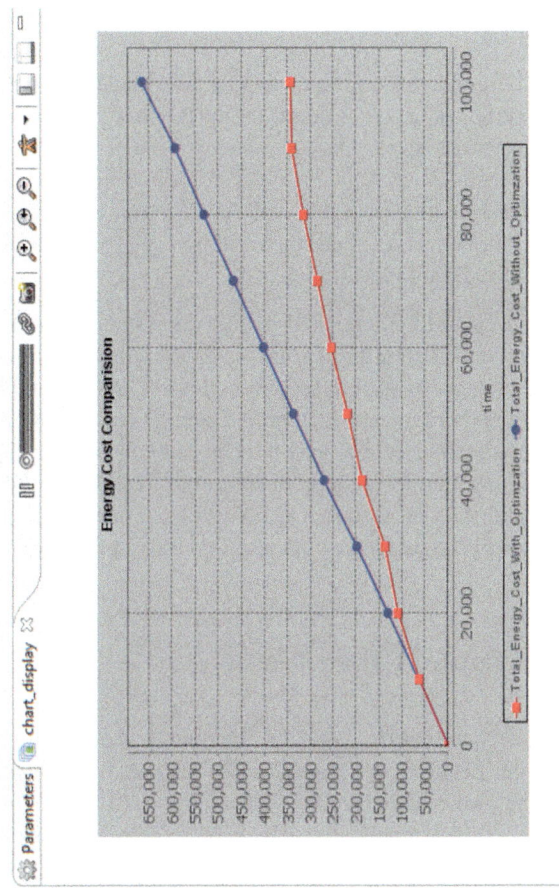

Fig. 5.4 Simulation results for energy cost comparisons (color figure online)

result of implementing UTES method to optimize transport, the energy consumptions savings are seen to be significant.

UTES provides users with a variety of interface options, and the underlying implementation and technology stack is quite diverse as a result. The UTES server back-end is written in Java and uses a variety of standard open source development libraries and frameworks for its implementation. The system is composed of a number of service modules, each providing specific functionality, coupled together.

In order to verify the performance of the proposed algorithm, a web-based navigation system was developed. The map information used is the open source map from the Baidu platform.

1. After the user has registered online, a unique user database is created in the Azure Platform. In the database, one table stores the user's current energy quota, another stores the travel logs.
2. The navigation website at present provides for three different transportation types: private car, public transport, and walk/cycle. Once the origin and destination and transport mode are selected, the total trip energy use will be provided immediately from the Baidu platform.

What levels of data would be generated by such a system? A rough calculation gives some guidance. Annual Beijing road traffic data are estimated to be about 12 terabytes (TB) per year, consisting of data from both GPS systems and loop detector data feeds. In 2015, the UN estimate for Beijing's population was 18 million people. Assume 5% of the population used the system twice daily on average, then total daily visits will be 1.8 million. Further assume uniform daytime distribution over about 10 h; the system would then need to meet about 60 access requests every second, although at peak hours this might rise to perhaps 300–500 requests every second [49].

5.7   Discussion: Future Urban Travel

Predicting what is likely to happen can help planning, not only in urban transport but also in energy and health. Even though our forecasts for transport and other areas are often wrong, we really have no other options. Consider the decision to investigate the construction of a new urban freeway or metro rail system. The investigation, design, and construction alone could take a decade, and the freeway or metro will have an operation life of many decades. Hence the planners have to try to predict its likely use many decades in the future if they are to assess the project's viability. Of course, in some cases, provision can be made for changing future use by stage construction, providing space for extra runways for an airport, or extra lanes for a freeway to be added if required in future.

Schäfer and Victor [40] forecast global vehicular travel out to 2050, on the basis of travel time and money budgets. Global air travel was expected to grow at a faster rate than car travel to accommodate growing passenger-km per capita within a daily travel time budget constraint. By 2050, they predicted that public transport and non-motorised modes would be marginal, given their slower speeds. Nevertheless, they anticipated global car travel to continue to grow strongly out to 2050, reaching about 45 trillion pass-km in that year, a prediction which receives some support from the Organization of the Petroleum Exporting Countries (OPEC) car ownership projections. For 2014, OPEC [36] gave an estimate of 1022 million cars globally (142 per 1000 population) and projected this to rise to 2167 million by 2040 (240 per 1000 population). Most of the increase was forecast to occur outside the OECD countries. Since fuel efficiency improvements were forecast to be insufficient to compensate for the increased vehicle usage, total passenger car energy use would continue to rise. Although urban car ownership is often lower than the national average in OECD cities, because of lower than average inner city resident car ownership, the opposite is usually true in non-OECD cities.

The global car ownership and travel forecasts both assume a business-as-usual world. But given the challenges faced by transport discussed in Chap. 1, it is difficult to see how climate mitigation goals for transport could be met unless actual travel and car ownership are a mere fraction of these forecasts. It then follows that the benefits obtained from each passenger-km of urban vehicular travel will need to be far higher than is the case today.

Chaoming Song and colleagues [41] studied the predictability of individual travel patterns by using the data traces left by mobile phone users. Mobile phone companies have data on the 'closest mobile tower each time the user uses his or her phone.' The researchers found 'a 93% potential predictability in user mobility'. They further claimed that, with the use of data mining algorithms, actual predictions about mobility could be obtained. Yu Zheng [52] of Microsoft Asia, has reviewed how trajectory analysis can be used to better understand the mobility patterns of not only people and vehicles, but even of wildlife, and has discussed how data collected from diverse sources can be mined to better understand how urban transport systems actually function.

Such data could be helpful to cities in moving toward sustainable urban transport systems. The results of such analyses are, of course, only valid for current transport patterns, which in turn reflect current transport policies. Nevertheless, these and similar approaches would allow frequent snapshots of travel to be made, and so allow tracking the changes occurring as a result of policies needed to support urban sustainability, and if found necessary, to alter policies in the direction of greater sustainability. They can also be used to track travel patterns and its changes at very local levels.

References

1. Aguiléra A, Guillot C, Rallet A (2012) Mobile ICTs and physical mobility: review and research agenda. Transp Res A 46:664–672
2. Alexander LP, González MC (2015) Assessing the impact of real-time ridesharing on urban traffic using mobile phone data. UrbComp'15, Sydney, Australia, August 10. http://humnetlab.mit.edu/wordpress/wp-content/uploads/2014/04/sig-alternate.pdf
3. Batty M (2013) Big data, smart cities and city planning. Dialogues Hum Geogr 3(3):274–279

4. Bodhani A (2012) Smart transport. Eng Technol 7:70–73
5. BP (2017) BP statistical review of world energy 2017. BP, London
6. Burns LD (2013) A vision of our transport future. Nature 497:181–182
7. Bruun E, Givoni M (2015) Six research routes to steer transport policy. Nature 523:29–31
8. Cameron I, Lyons TJ, Kenworthy JR (2004) Trends in vehicle kilometres of travel in world cities, 1960–1990: underlying drivers and policy responses. Transp Policy 11:287–298
9. Cohen-Blankshtain G, Rotem-Mindali O (2013) Key research themes on ICT and sustainable urban mobility. Int J Sustain Transp. https://doi.org/10.1080/15568318.2013.820994
10. Coughlan A (2012) The best medicine. New Scientist 215:38–41
11. Dal Fiore F, Mokhtarian PL, Salomon I et al (2014) "Nomads at last?" A set of perspectives on how mobile technology may affect travel. J Transp Geogr 41:97–106
12. Department for Transport (DtF) (2015) National travel survey: England 2014. DtF, London. (Also earlier surveys)
13. Funk JL (2015) IT and sustainability: new strategies for reducing carbon emissions and resource usage in transportation. Telecommun Policy 39:861–874
14. Gabrielli S, Maimon R (2013) Digital interventions for sustainable urban mobility: a pilot study. UbiComp'13, Zurich, Switzerland, September 8–12
15. Goel S (2016) Special issue on connected vehicles. IEEE Intell Transp Syst Mag 8:5–7
16. Gomes L (2016) When will Google's self-driving car really be ready? IEEE Spectr 53:13–14
17. Hannes E, Liu F, Vanhulsel M et al (2012) Tracking household routines using scheduling hypothesis embedded in skeletons. Transportmetrica 8(3):225–241. https://doi.org/10.1080/18 128602.2010.539418
18. Hodson H (2015) A city of numbers. New Scientist 225:22–23
19. Khan SM, Chowdhury M (2014) ITS for one of the most congested cities in the developing world—Dhaka, Bangladesh: challenges and potentials. IEEE Intell Transp Syst Mag 6:80–83
20. Lo AY (2016) Challenges to the development of carbon markets in China. Clim Pol 16(1):109–124
21. Luo X, Dong L, Dou Y et al (2017) Factor decomposition analysis and causal mechanism investigation on urban transport CO_2 emissions: comparative study on Shanghai and Tokyo. Energy Policy. https://doi.org/10.1016/j.enpol.2017.02.049
22. Lyons G (2015) Transport's digital age transition. J Transp Land Use 8(2):1–19. https://doi.org/10.5198/jtlu.2014.751
23. Mayer-Schönberger V, Cukier K (2014) Big data. Mariner Books, Boston
24. Menon A, Sinha R, Ediga D, Iyer S (2013) Implementation of internet of things in bus transport system of Singapore. Asian J Eng Res 1(4):8–17
25. Miller HJ (2013) Beyond sharing: cultivating cooperative transportation systems through geographic information science. J Transp Geogr 31:296–308
26. Moriarty P (1980) The potential for non-motorised transport in Melbourne. Proc 10th ARRB Conf Sydney 10(5):43–50
27. Moriarty P (2016) Reducing levels of urban passenger travel. Int J Sustain Transp 10(8):712–719
28. Moriarty P, Honnery D (2011) Rise and fall of the carbon civilisation. Springer, London
29. Moriarty P, Honnery D 2017. Reducing personal mobility for climate change mitigation. InW-Y Chen, T Suzuki and M. Lackner (Eds.-in-chief) Handbook of climate change mitigation and adaptation, 2nd Edition, Springer Science+Business Media: New York
30. Moriarty P, Honnery D (2017) Sustainable energy resources: prospects and policy, Chapter 1. In: Rasul MG et al (eds) Clean energy for sustainable development. Academic, London
31. Moriarty P, Kennedy D (2000) Will telework reduce travel? IS 2000 Conference. University of Woollongong, NSW
32. Moriarty P, Wang SJ (2014) Low-carbon cities: lifestyle changes are necessary. Energy Procedia 61:2289–2292
33. Moriarty P, Wang SJ (2015) Eco-efficiency indicators for urban transport. J Sustain Dev Energy Water Environ Syst 3(2):183–195

34. Newman P, Kenworthy J (1999) Sustainability and cities: overcoming automobile dependence. Island Press, Washington, DC
35. Nilles JM (1976) Telecommunications-transportation tradeoff: options for tomorrow. Wiley, New York
36. Organization of the Petroleum Exporting Countries (OPEC) (2016) World oil outlook 2016. OPEC, Vienna. (Also earlier editions)
37. Pan G, Qi G, Zhang W et al (2013) Trace analysis and mining for smart cities: issues, methods, and applications. IEEE Commun Mag 51(6):120–126
38. Pappalardo L, Rinzivillo S, Qu Z et al (2013) Understanding the patterns of car travel. Eur Phys J Special Topics 215:61–73
39. Peterson B (2014) The vision of automated driving: what is it good for? Answers from society with economical and organizational perspectives. Proceedings of the 44th European Microwave Conference, Rome, Italy, pp 1723–1729
40. Schäfer A, Victor D (2000) The future mobility of the world population. Transp Res A 34(3):171–205
41. Song C, Qu Z, Blumm N et al (2010) Limits of predictability in human mobility. Science 327(5968):1018–1021
42. Starkey R (2012) Personal carbon trading: a critical survey. Part 1: Equity. Ecol Econ 73:7–18
43. Statistics Bureau Japan (2015) Japan statistical yearbook 2016. Statistics Bureau, Tokyo. (Also earlier editions). http://www.stat.go.jp/english/data/nenkan/index.htm
44. Taniguchi E, Thompson RG, Yamada T (2016) New opportunities and challenges for city logistics. Transp Res Procedia 12:5–13
45. Tao S, Corcoran J, Mateo-Babiano I et al (2014) Exploring bus rapid transit passenger travel behaviour using big data. Appl Geogr 53:90–104
46. Thomopoulos N, Givoni M (2015) The autonomous car—a blessing or a curse for the future of low carbon mobility? An exploration of likely vs. desirable outcomes. Eur J Futures Res 3:14pp. https://doi.org/10.1007/s40309-015-0071-z
47. United Nations (UN) (2014) World urbanization prospects: the 2014 revision. https://esa.un.org/unpd/wup/CD-ROM/. Accessed 30 Nov 2016. Also the 2011 revision
48. Waldrop MM (2015) No drivers required. Nature 518:20–22
49. Wang SJ, Moriarty P, Ji YM et al (2015) A new approach for reducing urban transport energy. Energy Procedia 75:2910–2915. https://doi.org/10.1016/j.egypro.2015.07.588
50. Wikipedia (2017) Congestion pricing. https://en.wikipedia.org/wiki/Congestion_pricing. Accessed 22 May 2017
51. Wikipedia (2017) Floating car data. https://en.wikipedia.org/wiki/Floating_car_data. Accessed 14 Aug 2017
52. Zheng Y (2015) Trajectory data mining: an overview. ACM Trans Intell Syst Technol 6(3):1–41. https://www.microsoft.com/en-us/research/wp-content/uploads/2015/09/TrajectoryDataMining-tist-yuzheng.pdf
53. 龙瀛, 张宇, 崔承印 (2012) 利用公交刷卡数据分析北京职住关系和通勤出行. 地理学报 67(10):1339–1352

Chapter 6
Big Data for Urban Energy Reductions

6.1 Introduction

Globally, primary energy consumption has shown an extraordinary rise since 1900. In that year, energy use was only about 44 EJ, much of it fuel wood, but it rose to 101 EJ by 1950, and then to 574 EJ by 2014 [14, 22]. Because of the urgent necessity to seriously tackle climate change, the possibility of rising fossil fuel costs associated with depletion of easily-produced reserves, and uncertainties over both the technical potential of energy alternatives and their long lead times for implementation, global future energy use will most likely need to be curtailed [23, 25, 29]. Consequently, per capita energy use in many cities will have to fall, particularly in the high-energy consumption cities of the OECD [24]. A visual reminder of the high energy consumption of cities can be gained from satellite images of the Earth at night, with brightly lit city regions clearly visible.

Although big data could be an important aid to other areas of urban sustainability, as we discuss in other chapters, it will simply be *essential* for sustainable energy production [18], smart grids and energy efficiency [32]. In Chap. 1 we reviewed the possible approaches for a sustainable energy system in the future, concluding that RE will need to supply the greater part of our future energy needs. We also pointed out that only intermittent energy sources, largely wind and solar energy, have sufficient global technical potential to allow for the needed large expansion in global RE production. Yet the shift to non-fossil sources of energy is occurring much too slowly. When the first IPCC report was released in 1991, non-fossil fuel electricity accounted for 36.7% of global electricity generation, but by 2014, this figure had fallen to 33.2% [5]. Although global solar energy production is still growing exponentially, wind energy production growth is now growing only linearly.

Germany, a leading country in solar energy, has around 1.5 million photovoltaic (PV) cell installations, largely small-scale domestic roof-top installations, which together with medium and large scale solar PV farms, in 2014 generated as much as 7% of German electricity production [43]. The traditional electricity supply model involved a few large generating plants, with each typically having a capacity of 100 MW or more, supplying perhaps millions of households as well as many industrial and commercial customers. In contrast, with the rapid global expansion of rooftop PV units, electricity utilities are having to cope with millions of both producers and consumers. A further major difference is that, traditionally, utilities could plan their production, with baseload plants run continuously, and extra units brought in to match varying daily and seasonal demand. With intermittent solar and wind energy, production not only cannot be planned, but electricity output from these sources can be reliably estimated only a limited time into the future.

However, intermittent renewable electricity will be more costly than fossil fuel electricity, particularly if large amounts of storage prove necessary [25]. Storing wind or solar electricity will first require conversion of the excess electricity to another energy form such as chemical energy (as in batteries), hydrogen, methanol, or compressed air stored in underground caverns. It must then be re-converted back into electric power as needed, with further energy losses and costs. To encourage a major shift to these costlier non-carbon energy sources, a carbon tax (or some other means of pricing carbon) will therefore prove necessary. According to the IPCC [13], global temperatures increases (relative to pre-industrial temperatures) need to be kept below 1.5–2.0 °C if we are to avoid serious anthropogenic climate change. The modelled results of van Vuuren et al. [38] showed that to keep within this limit, a global carbon price as high as US$ 700–900 per tonne of carbon (or 200–250 US$ per tonne CO_2) would be needed by 2050, and would then need to be permanently maintained at this level.

Chapter 1 stressed both the need for more RE production, especially intermittent energy and the need to reduce overall energy consumption as fossil fuel use is progressively cut back. Although this chapter will concentrate on electricity, other energy carriers will also be important for urban areas in future, even if they are mainly ultimately sourced from renewable electricity.

The rest of this chapter consists of three parts. Section 6.2 discusses electricity production and the need for smart electricity grids in an era of multi-source intermittent energy producers. Sections 6.3–6.6 look at the demand side: how can big data help reduce energy use, from individual buildings through to the city as a whole? This brings us to an important general point: for many areas of urban sustainability, including energy conservation, big data is only an *enabling technology*. Other changes will need to occur for big data to realise its potential. A final section (Sect. 6.7) stresses the importance of an integrated view of urban energy consumption, necessary for reducing energy use and its consequent emissions.

6.2 Smart Grids: A Necessary Part of Sustainable Energy

In general, the purpose of smart grids is to enable 'a two-way flow of power and data between suppliers and consumers in order to facilitate the power flow optimization in terms of economic efficiency, reliability and sustainability' [10]. As with the terms 'big data' and 'smart cities', the phrase 'smart grids' means different things to different people. In some cases it is simply a wish list of the desirable properties any grid should have. Zora Kovacic and Mario Giampietro [15] have provided a list derived from the published literature of their—sometimes potentially contradictory—properties:

- Instantaneous matching of electricity supply with demand
- Reducing peak demand
- Enabling the transition to RE electricity
- Securing energy supply
- Reducing blackout frequency
- Increasing overall energy efficiency
- Guaranteeing electricity access for everyone
- Decentralising electricity generation.

Blumsack and Fernandez [4] summed up the vital importance of big data and its application to the smart grid as 'the ability to process and analyze large amounts of information.' But as they also pointed out, moving to smart grids will add to the complexity of the already complex conventional electric grids. They thus warned that there was no guarantee that blackouts would be less common or less severe with smart grids. As an illustration of this increased complexity, Andreottola et al. [1] have discussed how the introduction of large numbers of customers as small energy generators into the grid—as in Germany—brings with it a number of technical problems in addition to the random availability of intermittent RE.

In Sect. 1.2, we argued that attempts to 'green' fossil fuel-based grids by using CCS and/or geoengineering are unlikely to succeed, even assuming an adequate future supply of accessible fossil fuels. Several proposals have also been made for enabling the conventional approach of a few large generating units supplying many consumers to continue in an RE future, rather than decentralising generation as in the list above. An ambitious approach proposed by the Desertec project foresaw solar (and wind) energy produced from energy farms in the deserts of North Africa and the Middle–East sent thousands of km to supply electricity to European grids [25].

Even more visionary schemes have seen solar production from energy farms in all the worlds deserts, in both hemispheres, and in various time zones, connected in a worldwide grid to even out daily and seasonal supply fluctuations. Another method for ensuring a non-intermittent supply of solar electricity, originally proposed more than four decades ago, is to produce solar energy from an array of suitably positioned satellites made with large light-weight structures covered with PV cells, convert it to microwave energy, transmit this power to Earth-based receiving stations, then finally convert it back to electricity [20]. All these ambitious schemes would be extremely expensive, would take decades to become operational, and would raise many political and environmental problems and risks. For example, would countries be prepared to risk placing their electricity supply entirely in the hands of a distant country? At present, in marked contrast to oil, very little of the world's electricity production crosses national borders [5, 14].

Decentralising electricity generation, one of the properties of smart grids listed by Kovacic and Giampietro, would overcome at least some of these problems. One of the advantages of decentralisation is a reduction in the need for new transmission line capacity. In many mature industrial countries of the OECD, adding new capacity has proved very difficult, because of land constraints for new right-of-way and citizen opposition [3]. What is more, with the planned increase in RE electricity, transmission capacity will need to rise disproportionately, because it will have to be designed for peak production of intermittent electricity, not the average [26]. Local generation can help get around the problem of limits on transmission line capacity. So, of course, can reductions in electricity demand, discussed below.

Apart from demand management, discussed in Sect. 6.3, another obvious approach to dealing with the transition of electricity grids to intermittent RE is to provide energy storage. California has mandated that 1.32 GW of storage capacity be installed in the state by the year 2020 [16]. This energy storage can be at a large scale and provided by the grid utilities, as with the already common pumped water storage (in use since 1929 [41]), or the much rarer compressed air in underground caverns. If batteries are used, however, one possibility for these highly modular devices is for at least some of the capacity to be sited at domestic houses and other buildings; just as with PV rooftop cells for electricity production, some energy storage capacity would also be sited domestically.

Wang and colleagues [41] have discussed how such a system might work. Domestic energy bills have two important components: the wholesale cost of electricity and the investment costs for the distribution network. The latter cost can be reduced by the use of low voltage distribution networks. They developed a model to minimise the sum of both costs, in which domestic storage devices (in this case lithium ion batteries) are jointly operated by the household and the utility. They found that compared with the base case in which households operated the batteries solely to minimise their own power costs, joint operation, although it increased complexity, enables overall system costs to be reduced, which in turn would cut electricity costs for the household.

6.3 Urban Domestic Energy Consumption

Domestic energy use is typically around 20% of total primary energy use in OECD countries, with a similar percentage used in the commercial sector. In the US, about two-thirds of this domestic use is for space heating and cooling, and water heating [27]. Like urban transport, domestic electricity, gas and water bills often have high fixed costs, which discourage households from cutting energy or water use. Sometimes electricity can even be at zero cost to householders: Mikael Elinder and his colleagues [11] have discussed the case of a large apartment block in Sweden, where the residents were allowed unlimited electricity usage as part of their rent. As expected, when market-based charges were introduced for 800 of the apartments as an experiment, electricity use for these apartments dropped markedly compared with the remaining 1000 apartments still getting free electricity. As with urban transport, for reductions to occur, the structure of costs will need to change, with variable costs increased and fixed costs lowered.

Even worse, there are often *perverse incentives*, in that unit costs for energy are often lower for greater consumption. The fixed costs could be decreased anyhow in future, because the smart meters already adopted by grids in a number of countries can be read automatically, saving meter reading costs. Smart meters can supply information (such as real-time electricity prices) to both householders and electricity supply officials on both power use by individual electrical devices and the timing of power consumption. They can also determine what appliance is drawing power, as each has a different power consumption 'signature'. In contrast, 'time of use' meters can only differentiate between peak and off-peak usage [4].

Perhaps partly because of the present adverse cost structure, the installation of smart meters by itself has so far had a negligible measured impact on domestic electricity consumption [40]: a Swiss study found that potential electricity savings from smart meters were only 5–6% [46]. In the Swedish study discussed above, installing apartment level billing was found to reduce electricity use by about 25% compared with the apartments where no meters were installed [11]. Electricity prices for domestic consumers also vary greatly even for OECD countries, ranging in 2015 from under 10 US cents to over 30 cents per kWh, and industry prices for electricity (and natural gas) are often less than half that for domestic consumers [14].

More generally, researchers have found that merely providing more information to the public on energy consumption, or why energy savings are needed, has not been very successful [9, 35]. As UK energy researcher Steve Sorrell [37] has put it: 'With energy costs being small, largely invisible and poorly understood in most relevant situations, the more common situation is unreflective, habitual energy consumption in which energy costs are secondary to other factors such as convenience and symbolism and where energy-using behaviours exhibit considerable inertia.' Supporting policies are needed to supplement more detailed information.

The results of interviews with UK householders also argue against viewing the domestic energy conservation problem as a simple information gap [35]. Not only does the information provided need to fit in with the householder's personal circumstances, but it has to be understood by the householder. Although relating energy savings to lower energy bills is one way of doing this, this self-interested approach runs the risk of crowding out deeper changes, and in any case is self-defeating if changes in energy prices render the money savings trivial or, worse, negative—the same is true for petrol savings in private transport [28], where real costs have fluctuated greatly in recent decades.

Nevertheless, some researchers feel confident that better application of various social psychological principles can leverage the energy information provided into much more significant energy savings. Various approaches that have been tried to promote such pro-environmental behaviour (PEB) include: goal setting, where an energy reduction target is set; comparison of individual consumption with others; and improving the timing and targeting of the information [46]. Tom Hargreaves and others [12] found that although householders' knowledge about details of their electricity use increased with increased information provision, reductions in use beyond a modest level proved difficult to achieve. The authors suggested that the lack of official support for electricity reductions was largely responsible.

But what all these accounts miss are the profound changes that the combination of smart grids and the dominance of intermittent RE for electricity supply will bring in the future. Today (as we saw for Germany), and even more so in the future, many householders, and owners of non-residential buildings are—or will be—both consumers of electricity *and* producers (mainly through roof top PV cell arrays) [4]; they can be regarded as *prosumers*. As active producers, they will have far more interest in energy production and consumption than is presently the case, particularly in the unit prices for electricity they are generating.

In Sect. 6.2, the 'instantaneous matching of electricity supply with demand' was listed as a (desirable) property of smart grids. We can only guess as to how this could be achieved in an intermittent energy future, but load management using big data could be the cheapest approach. One possibility is that domestic electricity consumers will receive daily renewable energy generation forecasts, just as today they receive weather forecasts for up to a week ahead. (In fact, the two types of forecast are closely related, given the dependence of RE on wind, insolation levels and cloud cover). These forecasts may be accompanied by forecasts of the unit costs of such electricity. Households could then save on electricity costs—which are likely to be much higher in real terms than today's—by shifting some activities to times predicted to have low unit electricity costs because of plentiful wind and/or insolation. Possible activities would include clothes washing (and clothes drying, where, as in the past, wind energy could be used directly by use of clothes lines), household vacuuming, recharging any domestic storage or electric vehicle batteries, and so on. Just as today we consult official weather forecasts in order to decide whether to take a raincoat and umbrella, in future we will attend to energy supply and cost forecasting to decide upon the timing of activities. Householders will come to accept the resulting unit price variability in the same way as weather variability is accepted, and will realise that electricity on demand at any time—at least at affordable prices—is a relic of the fossil fuel age, and is no longer a feasible option in an era of intermittent RE.

Eventually, households will come to know the seasonal variations in energy supply for their region and thus how energy costs vary over time, and will use this knowledge to guide their (new) daily routines. (In some cases seasonal energy supply and demand will be a good match: in warm regions air-conditioning will be in greatest demand precisely when insolation, and thus supply from solar energy, is also at its peak.) It may be as simple as less electric power being available in winter if RE is mainly produced from solar energy and none at all after sunset. For solar thermal electric systems, some storage of power as thermal energy is possible after sunset, so unit price changes should be more gradual. In contrast to today, night time electricity prices could well be higher than daytime prices. If hydro is important, it will depend on the annual rainfall variation in the catchment area. For wind power, there may well be daily as well as seasonal variations.

We have to consider not only households but also larger electricity consumers, such as commercial buildings and industries. Some industrial processes, such as aluminium smelting, cannot tolerate any interruption to their power supplies and will have to pay a premium for this uninterruptible service. But commercial offices and some industrial operations will be in a similar position to households, and can time shift at least some of their energy use.

How relevant is the above discussion to the cities low- and middle-income countries? For a start, grid connection rates are usually very high in middle-income countries. It is 100% in China and already nearly 80% in India, and approaching 100% in most Latin American countries. Although connection rates are usually much lower in tropical African countries—sometimes less than 10%, connection rates in cities are far higher [45]. Given that these countries are usually well-situated for PV energy, smart grids should be just as relevant as in OECD cities.

6.4 Smart Grids and Electric Vehicle Charging

Many commentators see battery electric vehicles (EVs) as the future for transport, as a means of both reducing urban pollution (both air and noise) and transport oil dependency. Further, if the electricity was supplied by RE, transport GHGs would also be greatly reduced. Although by mid-2016, over 11 million hybrid EVs and over 1.5 million plug-in EVs had been sold worldwide [44], their introduction over the past two decades has been slow, given that the total global car fleet in 2014 numbered over 1000 million [33]. In follows that their impact on electric grids has so far been minimal, particularly since hybrid EVs run on conventional, mainly oil-based fuels, but with an electric drive train. Although China has tens of millions of electric bicycles, their grid impact is again negligible.

All this could change in the future if full battery EVs were to replace conventionally-fuelled cars, buses, and even some freight vehicles on roads. One estimate is that the global car fleet will rise to 2167 million by 2040, mainly by the continuation of the rapid growth in the presently low-car ownership non-OECD countries [33]. Even with existing car ownership levels, household domestic electricity use in typical OECD cities would be roughly doubled by the charging needs of these vehicles [34]. Further, charging would usually be done during the evening peak period for domestic energy use, which could overload existing grids. Electric grids would then face two problems: integrating intermittent RE as the dominant electricity source and managing the power load from EV charging. Smarter grids would be essential for resolving these problems.

One strategy discussed for overcoming these problems is to connect EVs to the grid so that their batteries could be used for 'vehicle-to-grid' (V2G) energy storage. The vehicle owners, whether private owners or fleet operators, could then sell electricity to the grid when the demand for electric power exceeded the available supply. Today, private vehicles are parked for about 95% of a 24 h day [31]. Their connection to the grid during this idle time would enable transfer of electrical energy to and from the vehicle's battery pack. In addition, V2G technology could help regulate voltage and frequency, and provide spinning reserve.

Nevertheless, we think that global EV numbers will never rise to anywhere near even existing levels of vehicles. Their power demands would simply be too high in a world which will need to significantly cut global energy consumption, if fossil fuels, the present energy mainstay, have to be phased out. Further, electric public transport can (and already does in many cities) also deliver urban air pollution benefits and GHG reduction benefits, given that it is several times as energy efficient as private travel. But with the high cost of pure EVs compared with conventional vehicles—largely because of the cost of their battery packs—car sharing could become common, with the vehicles owned by 'Mobility Service Providers' [34]. Public transport and taxis can be considered to already operate in this mode. Shared vehicles would be driven many more annual km than today's vehicles (just as today's taxis and public transport vehicles are), and consequently would be parked less often, especially during daytime hours. (Fully automated vehicles could also encourage car-sharing, but as explained in Chap. 5, we consider their widespread introduction unlikely.) With night-time battery recharging, vehicles would be drawing grid power at a time when no solar power from PV cells was available, rather than supplying it to the grid, and the advantages of V2G would be lost [30]. Since V2G can only work if the car fleet stands wastefully idle most of the day, it will at best be a transition technology. Unlike household energy storage discussed in Sect. 6.2, V2G will probably have at best a minor role to play; future ownership of vehicles is uncertain, but people will still live in buildings.

6.5   Smart Buildings

Smart houses (and smart buildings generally) are a necessary complement to smart grids. Similarly, smart electricity meters are needed for buildings to be considered 'smart' [17]. Here we will focus on smart houses and their potential role in reducing energy use and GHG emissions. In the coming era of intermittent energy supply, changing the *timing* of energy use may be as important for sustainability as reducing energy consumption.

We have already discussed the active role of householders in shifting energy-intensive *activities* to periods of high power availability/lower prices. Here we consider another important means for energy demand management: automated electric power demand shifting, an essential feature of smart buildings. Domestic energy-intensive appliances are of two types: those such as washing machines and dryers, and dishwashers, which are only operated intermittently, and appliances such as freezers, refrigerators, and domestic hot water services, which are run continuously. With Internet-connected smart appliances [39], the first group could be programmed to operate for minimum energy cost as an alternative to active householder operation. With continuously-run smart appliances, it would be possible to switch them off for short periods at times of high prices without any noticeable ill-effects [7].

Similarly, refrigerators and freezers could be run colder, and water heating units hotter, for short periods when electricity prices were low.

Another significant potential use for big data in buildings is for predictive maintenance. In Italy, software researchers analysed the year 2015 temperature, humidity, and electric power use data from the air conditioning and ventilation unit at a hospital [2]. They used data from the first half of 2015 to train the algorithms used, then applied it to the data for the second half. The algorithms were largely successful in predicting faults that did occur in the second half of 2015 and had only a 5% rate of false positives. Heating, ventilation and air conditioning represent about three-quarters of the total energy consumption of buildings in Europe [21]; hence any reduction in this load is significant. Already, in many Finnish apartment blocks, sensors have been installed to monitor temperature, humidity and air pressure in the apartments, and this monitoring enables the attainment of optimal conditions at reduced energy cost [2]. Predictive maintenance based on machine learning is also already proving its worth for industry as well.

Not only is the energy used by appliances in buildings important, but so is the energy used to construct and maintain the buildings over their service life. For example, replacing concrete and steel in buildings by timber can cut down on energy costs and GHG emissions. At the end of its useful construction life, the timber can then be combusted for energy, giving further GHG reductions [26]. Furthermore, attention to building design can not only save construction energy but even cut down on appliance energy use. Christian Calvillo and colleagues [6] have recently reviewed the literature on smart cities from such an integrated energy use viewpoint. They pointed out that if geographic information systems and 3-D modelling are used in building design, advantage could be taken of the local terrain to minimise occupant energy use, such as orientation for passive solar heating and light, and cross-ventilation from winds. Advantage can even be taken of sloping sites by partly burying the building to provide insulation and save on heating energy.

6.6 An Integrated View of Urban Energy Use

So far in this chapter, we have looked at several components of the urban energy system, while Chap. 5 discussed the energy aspects of transport. But just as climate scientists look at Earth energy flows [29], increasingly urban planners will have to look at urban energy flows as a system, if the aim is large urban energy reductions. All energy use in a city—whether from appliances, factories, power stations, street lighting, or vehicles—finishes up as waste heat. Here we look at how urban energy use could be viewed as a whole, so that, for example, the waste heat from one use could be utilised as an energy input to another. Energy cannot be created or destroyed, but it can be reused for purposes that only need lower-quality energy, just as the waste heat from vehicle engines is used to warm the vehicle interior.

The EU is currently funding research which aims to do just this [19]. The EU study is examining the ways that waste heat from such varied sources as underground railway stations, power sub-stations, waste incinerators, and even the heat from the wastewater of domestic baths and washing machines could be harvested and distributed to nearby residences. The potential is huge: a Danish study found that waste heat generated across Europe is more than enough to heat all its buildings [19]. With a city-wide system of temperature sensors, it would be possible to determine where large enough waste heat sources—and their heat output variation over time—were located to make their utilisation feasible, and, perhaps, in summer months, where waste heat generation should be reduced for the comfort of local residents.

Waste heat is also directly produced by thermal power stations—even for the most efficient power stations, more than half the input energy can be lost as heat. In principle, this waste heat could be used in district heating schemes, for hot water in all cities, and for space heating for cities in temperate and cold climates. At present, most power stations are designed to maximise electricity output for a given energy input. But if the waste heat is seen as an energy resource, rather than merely waste, then conventional power station efficiency is less important. For combined heat and power, the power stations would need to be located in the district because waste heat, cannot be feasibly transmitted by pipe more than a few km without excessive heat losses, unlike electricity, where transmission distances of hundreds of km are common. An apparent disadvantage is that such local power generating units would have much lower power output and electric conversion efficiency than larger units. This need not matter, as the ultimate test of efficiency is whether or not the combined heat and power (CHP) system has lower input energy needs and lower GHG emissions for satisfying a given set of energy services, compared with the present conventional systems, which use natural gas for heating, as well as electricity for its usual applications.

But such approaches may not be a sustainable solution, even if useful during the transition to a more environmentally sustainable urban future. As the world reduces its dependence on fossil fuels, the present dominance of thermal power stations will wane. Only renewable biomass power generation will have waste heat available for CHP systems. Waste incinerators—which can also be used for power generation—also produce waste heat, but with less waste production and more recycling expected in a sustainable city, this energy source might only be temporary.

Another possible source of waste heat for CHP schemes is from the combustion of methane tapped from landfills, which is in effect mining the energy from past urban waste. Use of landfill gas has two advantages for climate mitigation: it both substitutes biogenic methane for fossil fuels and also prevents emissions to the atmosphere of methane, an effective greenhouse gas. But, like methane from urban sewage plants, it can only ever be a minor urban energy source, and more recycling would see its potential decline over time. On the other hand, more use of timber for construction would eventually increase bioenergy use when burnt after its useful life as a construction material has ended.

An integrated view of urban energy use can also help avoid cases of sub-optimisation. For example, higher urban residential densities are often advocated as a means of reducing urban travel, and with it transport energy use and emissions (see Chap. 5). This approach is doubtless effective at very high levels of urban density, as indicated by the low per capita transport energy used in high-density cities like Tokyo and Hong Kong. The monetary and energy costs of urban infrastructure per building served are also lower because given lengths for roads, water pipes, and cables are shared by more users. And it also helps with district heating schemes, since it means more potential users within a given radius. But, as discussed in Chap. 5, there are other ways of managing urban travel demand which do not require the extensive modification of the presently low-density cities common in North America and Australasia.

The daytime urban heat island (UHI) effect is partly the result of the high levels of waste heat release per km^2 and the reduction in evapo-transpiration surface, both characteristics of cities. However, the energy savings discussed above from promoting higher densities may be at the expense of other areas of urban energy use—it may be a sub-optimal solution. High density living translates into high intensities of energy use per hectare, and high proportions of impermeable surfaces. All these features exacerbate the UHI effect and the need for air-conditioning. It also means that the ability to use *passive* solar energy—for heating, cooling, and providing natural light and ventilation to buildings—is curtailed. Suitable building and roof surface areas per capita for PV cell installation are also reduced. Higher densities also mean less vegetation, and the trees in urban parks, through evapotranspiration, can help cool the surrounding areas.

Unlike daytime UHI, most of the temperature rise from night time UHI results from the structural morphology of cities, and its impact on the release of stored solar energy from the surfaces of buildings at night [36]. Night time UHI can thus be reduced by changes to building heights, building surface materials—and their ability to both store and release heat—and 'sky view', which is a function of the building footprint. Since building density and building material properties also are essential for determining heating and cooling energy for buildings, many sensor measurements are needed for each urban location and time, to calculate the optimal path which would minimise total energy needs for the city. In existing fully built-up areas, major changes would be difficult but could be readily incorporated into newly developed urban areas, particularly in the cities of countries undergoing rapid urbanisation.

Finally, any new industrial plants, waste depots, and major transport routes should be sited to both minimise pollution for residents (whether from air, noise, or smell) and total transport distances, perhaps by co-location of complementary industries, as suggested by industrial ecology [42].

6.7   Discussion: Energy and Urban Sustainability

Caution is still needed to make sure that even an integrated approach to urban energy use as discussed in Sect. 6.6 does not shortchange other aspects of an ecologically sustainable and liveable city. Reducing total energy use will surely be necessary for climate change mitigation, but it must occur in such a way that is not at the expense of other urgent concerns such as environmental amenity, health, and equality. Just as not all big data applications will necessarily save urban energy (as shown for

AVs), so not all options for real urban energy savings can be taken up, because of conflicts with urban well-being. The aim clearly must, therefore, be to derive the greatest amount of *energy services* from each unit of primary energy used and to maximise use of passive solar energy.

Finally, the energy requirements of big data itself need to be monitored [32]. Some researchers have argued that the electric power costs of Internet transmissions may themselves be significant. According to Vlad Coroama and Lorenz Hilty [8], published estimates for energy intensity over the decade 2000–2010 have been in the range 0.0064 to 136 kWh per gigabyte of data transmitted, although estimates have declined over time. Another reason for this four orders of magnitude variation found was whether or not the power consumption of end devices was included. If, as seems reasonable, it must be included for a full assessment, the energy costs are significant and need to be reduced.

References

1. Andreottola G, Borghetti A, Di Tonno C et al (2015) Energy systems for smart cities. https://www.researchgate.net/profile/Gabriella_Trombino/publication/283085941_Energy_Systems_for_Smart_Cities_-_White_Papers_from_the_IEEE_Smart_Cities_Inaugural_Workshop_December_2014_in_Trento_Italy/links/5629fa7f08ae04c2aeb14bb6.pdf
2. Baraniuk C (2017) Buildings predict their own faults. New Scientist 233:12
3. Buijs P, Bekaert D, Cole S et al (2011) Transmission investment problems: going beyond standard solutions. Energy Policy 39(3):1794–1801
4. Blumsack S, Fernandez A (2012) Ready or not, here comes the smart grid! Energy 37:61–68
5. BP (2017) BP statistical review of world energy 2017. BP, London
6. Calvillo CF, Sánchez-Miralles A, Villar J (2016) Energy management and planning in smart cities. Renew Sust Energ Rev 55:273–287
7. Cook DJ (2012) How smart is your home? Science 335:1579–1581
8. Coroama VC, Hilty LM (2014) Assessing Internet energy intensity: a review of methods and results. Environ Impact Assess Rev 45:63–68
9. Delmas MA, Fischlein M, Asensio OI (2013) Information strategies and energy conservation behavior: a meta-analysis of experimental studies from 1975 to 2012. Energy Policy 61:729–739
10. Diamantoulakis PD, Kapinas VM, Karagiannidis GK (2015) Big data analytics for dynamic energy management in smart grids. Big Data Res 2(3):94–101
11. Elinder M, Escobar S, Ingel Petréa I (2017) Consequences of a price incentive on free riding and electric energy consumption. Proc Natl Acad Sci U S A 114:3091–3096
12. Hargreaves T, Nye M, Burgess J (2013) Keeping energy visible? Exploring how householders interact with feedback from smart energy monitors in the longer term. Energy Policy 52:126–134
13. Intergovernmental Panel on Climate Change (IPCC) (2015) Climate change 2014: synthesis report. Cambridge University Press, Cambridge, UK
14. International Energy Agency (IEA) (2016) Key world energy statistics 2016. IEA/OECD, Paris
15. Kovacic Z, Giampietro M (2015) Empty promises or promising futures? The case of smart grids. Energy 93:67–74
16. Lemmon JP (2015) Reimagine fuel cells. Nature 525:447–449

17. Louis J-N, Caló A, Pongrácz E (2014) Smart houses for energy efficiency and carbon dioxide emission reduction. Energy 2014: The Fourth international conference on smart grids, green communications and IT energy-aware technologies, pp 44–50. Available at https://www.researchgate.net/profile/Antonio_Calo/publication/261795426_Smart_Houses_for_Energy_Efficiency_and_Carbon_Dioxide_Emission_Reduction/links/02e7e5357d59edb3d0000000.pdf
18. Lund H, Andersen AN, Østergaard PA et al (2012) From electricity smart grids to smart energy systems—a market operation based approach and understanding. Energy 42:96–102
19. Marks P (2013) Can't stand the heat? Use it. New Scientist 220:24
20. Marks P (2016) Star power. New Scientist 229:38–41
21. Moreno MV, Dufour L, Skarmeta AF et al (2016) Big data: the key to energy efficiency in smart buildings. Soft Comput 20:1749–1762
22. Moriarty P, Honnery D (2011) Rise and fall of the carbon civilisation. Springer, London
23. Moriarty P, Honnery D (2012) Preparing for a low-energy future. Futures 44:883–892
24. Moriarty P, Honnery D (2015) Future cities in a warming world. Futures 66:45–53
25. Moriarty P, Honnery D (2016) Can renewable energy power the future? Energy Policy 93:3–7
26. Moriarty P, Honnery D (2016) Review: Assessing the climate mitigation potential of biomass. AIMS Energy J 5(1):20–38
27. Moriarty P, Honnery D 2017. Non-technical factors in household energy conservation. In W.-Y. Chen, T. Suzuki and M. Lackner (Eds.-in-chief) Handbook of climate change mitigation and adaptation, 2nd Edition. Springer Science+Business Media: New York
28. Moriarty P, Honnery D 2017. Reducing personal mobility for climate change mitigation. In W.-Y. Chen, T. Suzuki and M. Lackner (Eds.-in-chief) 'Handbook of climate change mitigation and adaptation', 2nd Edition. Springer Science+Business Media New York. (DOI https://doi.org/10.1007/978-1-4614-6431-0_73-1)
29. Moriarty P, Honnery D (2017) In: Rasul MG et al (eds) Clean energy for sustainable development. Academic, London
30. Moriarty P, Wang SJ (2017) Can electric vehicles deliver energy and carbon reductions? Energy Procedia. https://doi.org/10.1016/j.egypro.2017.03.713
31. Mwasilu F, Justo JJ, Kim E-K et al (2014) Electric vehicles and smart grid interaction: a review on vehicle to grid and renewable energy sources integration. Renew Sust Energ Rev 34:501–516
32. O'Grady M, O'Hare G (2012) How smart is your city? Science 335:1581–1582
33. Organization of the Petroleum Exporting Countries (OPEC) (2016) World oil outlook 2016. OPEC, Vienna. (Also earlier editions)
34. Schaeffer GJ, Belmans RJM (2011) Smartgrids- a key step to energy efficient cities of the future. IEEE Power and Energy Society General Meeting. Available at http://ieeexplore.ieee.org/xpl/login.jsp?tp=&arnumber=6039255&url=http%3A%2F%2Fieeexplore.ieee.org%2Fxpls%2Fabs_all.jsp%3Farnumber%3D6039255
35. Simcock N, MacGregor S, Catney P (2014) Factors influencing perceptions of domestic energy information: content, source and process. Energy Policy 65:455–464
36. Sobstyl JM, Emig T, Abdolhosseini Qomi MJ et al (2017) Role of structural morphology in urban heat islands at night time. Available at https://arxiv.org/pdf/1705.00504.pdf
37. Sorrell S (2015) Reducing energy demand: a review of issues, challenges and approaches. Renew Sust Energ Rev 47:74–82
38. van Vuuren DP, Stehfest E, Elzen MG, Kram T, Vliet JV, Deetman S et al (2011) RCP2.6: exploring the possibility to keep global mean temperature increase below 2 °C. Clim Chang 109:95–116
39. Vlot MC, Knigge JD, Slootweg JG (2013) Economical regulation power through load shifting with smart energy appliances. IEEE Trans Smart Grid 4(3):1705–1712
40. Wang SJ, Moriarty P (2016) Strategies for household energy conservation. Energy Procedia 105:2996–3002
41. Wang Z, Gu C, Li F et al (2013) Active demand response using shared energy storage for household energy management. IEEE Trans Smart Grid 4(4):1888–1897

42. Weisz H, Suh S, Graedel TE (2015) Industrial ecology: the role of manufactured capital in sustainability, vol 112. Proc Natl Acad Sci U S A, pp 6260–6264
43. Wikipedia (2017) Solar power in Germany. Available at https://en.wikipedia.org/wiki/Solar_power_in_Germany
44. Wikipedia (2017) Electric vehicle. Available at https://en.wikipedia.org/wiki/Electric_vehicle
45. World Bank (2017) Access to electricity (% of population). http://data.worldbank.org/indicator/EG.ELC.ACCS.ZS?view=chart. Accessed 21 Mar 2017
46. Zimberi G, Gautschi F (2014) Smart meters as an eco-feedback technology to motivate the reduction of electric energy consumption. Fact Sheet. Available at https://files.ifi.uzh.ch/hilty/t/examples/IuN/Smart_Meters_as_an_Eco_Feedback_Technology_Zimberi_Gautschi.pdf

Chapter 7
Big Data for Urban Health and Well-Being

7.1 Introduction

The global healthcare system is already under stress from the rise in population longevity and healthcare costs, as discussed in Chap. 2. But further threats to health could come from on-going climate changes and the rise (or re-emergence) of new diseases. Ford et al. [35] have discussed the potential role for the judicious use of big data—alongside the continued use of 'small data'—for managing the risks from climate change. Big data sets can be used to provide early warning for disease outbreaks or natural disasters, for example by using social media for early warnings of storms or earthquakes. But social media data can also be mined so that emergency relief agencies can determine which areas were the most severely damaged. Also of interest is determining how the general public respond to official warnings for hazards, which can be used for their improvement. Mobile phone records and social media could be used to track the movement of people in the affected area both before and after a severe storm, for example, to gauge which type of warnings were the most effective [35].

Chapter 2 discussed the various health and well-being problems facing urban residents and showed that they could be very different for residents in high-, middle-, or low-income cities. In this chapter, we first discuss the potential for big data to improve health and well-being in OECD countries and follow with examples of its applications that have been implemented, are being trialled, or are planned. Next, we discuss the Quantified Self (QS) movement in OECD countries and the possible role in future healthcare of QS or similar trends. We then examine the potential role of big data approaches in industrialising countries—and give some examples of applications that have already been implemented or trialled—stressing that the barriers to its successful implementation may be greater than in OECD countries. We finish with a case study, which could serve as an example of QS-type approaches, and then discuss how the data generated could be used in medical research.

© Springer International Publishing AG, part of Springer Nature 2018   119
S. J. Wang and P. Moriarty, *Big Data for Urban Sustainability*,
https://doi.org/10.1007/978-3-319-73610-5_7

7.2   The (Contested) Potential for Big Data in OECD Healthcare

The future role of big data in health care is contested. Some observers, while aware of teething problems, feel that big data will transform both medicine and healthcare in OECD countries as well as lowering costs [9, 27, 57, 63, 70, 75, 80]. McKinsey and Company, in a report with a health industry focus, have even written about 'The 'big data' revolution in healthcare', and claimed that a tipping point is near [39]. Similarly, a *Nature* editorial [5] stressed not only the potential for big data but also the problems that must be overcome, such as the shortage of medical data scientists needed to realise its healthcare potential. The editorial pointed out the huge volumes of medically relevant data likely to be available in future, with an estimate that 'clinical data from a single individual will generate 0.4 terabytes of information per lifetime, genomics data around 6 terabytes and additional data, 1,100 terabytes.' Like other researchers (e.g. [49, 53, 85]), the editorial stressed the danger of large data sets producing spurious correlations. As Muin Khoury and John Ioannidis [49] put it: 'Paradoxically, the proportion of false alarms among all proposed 'findings' may increase when one can measure more things'. Given that Ioannidis [47] has already argued that most medical research findings, even in top quality journals, are likely to be in error, this is a serious problem.

Nitesh Chawla and Darcy Davis [22] have pointed that despite the remarkable advances in genomics and collection of huge volumes of medical data, we still lack a systems view of various diseases. They advocated a personalised healthcare approach, but not necessarily one based on genomics, and stressed that *context* is crucial for understanding diseases and their treatment and that diseases result 'from an interaction between genetic, molecular, environmental, and lifestyle factors.' They therefore argued that personalising healthcare requires developing a disease risk profile for each patient, not only by using that patient's (electronic) medical record but also by looking for 'similarities of that patient to millions of other patients.' This approach to personalised healthcare would thus be based on data mining [44].

As shown in Sect. 2.1, a key concern for future healthcare provision in all countries is its rising costs; costs are even rising as a share of GDP. For the US, one estimate for the early 2010s was $ 2.6 trillion, 75% of which was for chronic disease management [77]. One important possible justification for using the huge data volumes generated is thus the potential for cost savings. In 2005, researchers at the RAND Corporation estimated that, *provided certain conditions were met*, 'rapid adoption of health information technology (IT) could save the United States more than $81 billion annually'. However, 7 years later, according to an update paper by RAND researchers Arthur Kellermann and Spencer Jones [48], actual health expenditures in the US had risen by $800 billion. They argued that the reason why health costs rose instead of falling was that three important conditions in the original 2005 study were not met. In brief, they argued that in the US, modern IT health systems:

- were not 'interconnected and interoperable'
- were not adopted as widely as they were in Europe, although their use was growing in the US

• were not used effectively, possibly because the available systems were not easy for health care practitioners to use.

Overall, they identified the core barrier to cost reductions in the US health sector through the use of new information technology as the fee-for-service model. Their claim was that unless this model was phased out there would be little incentive for the health care industry to use data-intensive technology to reduce medical costs; it could even be used to increase them. Their findings reinforce the message learned in other sectors of the economy, that even where application of big data is appropriate, it is a necessary but not sufficient condition for promoting more sustainable—and less costly—practices. Another way in which big data might reduce costs is in the detection and prevention of fraud and abuse in medical payments [16].

Others are more cautious about the potential for big data to enhance public health, even if the problems just discussed in a US context can be overcome. Gina Neff [62] provocatively entitled her 2013 article *'Why big data won't cure us'*. Her arguments paralleled those of Kellermann and Jones, emphasising, like they did, that the use of big data (or of modern IT in general) in healthcare, still had several obstacles to overcome before it could prove its worth, including how to guarantee the privacy of the vast volumes of data collected on patients. Nevertheless, as a 2016 US report on the subject found, 'medical error is the third leading cause of death in the US' [76]. Many of these deaths would have been the result of incorrect medical diagnoses, partly caused in turn by doctors being unable to keep up with the vast literature on medical advances. Big data approaches have the potential to avoid such deadly misdiagnoses. For this and other reasons, there has been an exponential rise in health-related big data research [29].

A very different approach to healthcare was provided by Jocalyn Clark [23]. She argued that healthcare is increasingly being 'medicalised', by which term she meant that far too much emphasis is placed on short-term approaches, with their priority for medical interventions, new vaccines and drugs and new equipment (the '3Ds' of doctors, drugs, and devices). (The views of Clark are echoed in an urban health context by Christopher Dye [30], and Melanie Swan [83], discussed in Sect. 7.2.2). Clark stressed that even in recent times, 'only about 10–43% of population health is thought to be attributable to healthcare'. The rest is the result of actions taken *outside* the health sector proper, such as for income and its distribution, education and pollution control.

It follows that for big data to be most effective for urban health, it must be able to address both the problems of conventional health delivery *and* facilitate actions outside the health sector which are vital to health. A good example, discussed in Chap. 5, would be the use of big data to encourage a major shift to non-motorised modes for urban travel. Not only would the exercise provide direct health benefits, but the resulting reductions in motorised travel would lower both air and noise pollution, resulting in further health benefits through a benign circle [60]. Nevertheless, many of the proposed applications of big data in healthcare discussed below have the side benefit of reducing trips to hospitals or clinics; they therefore cut the energy use and GHG and air pollution emissions from these trips. From another angle, directly limiting air travel could help slow the spread of vector-borne and contagious diseases [86].

7.2.1 Examples of Big Data Applications in OECD Healthcare and Well-Being

We have already discussed (see Chap. 2) the growing share of older people in the global population, but particularly in the heavily urbanised countries of the OECD. For the aged, continuous monitoring of health can be important. Below is a list of possible applications of big data in healthcare that are being developed or are already being used, the first three of them for the elderly. It should be noted that these applications have usually been developed in isolation, to solve a particular health problem, such as testing for Parkinson's disease. As such, these isolated uses of big data will be less effective than would a more systematic approach to health [32]. Even so, they typically result in fewer visits needed to a clinic or physician, and so can save on both patient's time and costs. They could also improve the quality of life for the elderly by prolonging independent living.

- As an example of how embedded sensors can aid the elderly, French researchers have described a device which can detect a person falling, send this information to a monitoring station and enable the accurate location of the fallen person using a GPS system [20].
- A smart phone application (called 'Indirect Wayfinding') has been developed to help elderly people, possibly with mild dementia, find their way in unfamiliar surroundings [2]. The app allows customisation for the individual concerned and allows the caregiver to use a web portal to suggest possible destinations etc. based on the GPS location of the elderly individual. Another smartphone app called iTrem 'uses the phone's built-in accelerometer to monitor a person's body tremors for Parkinson's disease'. The app could not only lower medical costs by dispensing with costly medical tests but can also allow health professionals to assess the disability remotely [56]. In a further application of Artificial Intelligence for Parkinson's disease, University of London researchers used deep learning to enable their smartphone app to distinguish between useful and spurious data, such as that resulting from the phone being accidentally knocked. It will not only allow sufferers to do the tests at home, but pooled data will also help determine the influence of lifestyle factors on the symptoms [71], adding to our knowledge of its causes and possible means of prevention. More knowledge about the disease in general and for each patient can be gathered with fewer healthcare visits.
- As another example of how data collected for one purpose can be used for an entirely different purpose, in the UK a system ('Howz') is being tested by the National Health Service (NHS) that plugs into electricity meters [73]. Just as smart meters do, it collects information on which household appliances are being

used at any given time and for how long they are switched on. By comparing use with the average of a few days, the system can flag any change in patterns—for example, an oven being left on too long. It can then contact a designated person. Howz is presently being trialled for monitoring people who have mild dementia living with a carer. In a related example, as reported by Cook [26], one study even claimed 'to find a link between changes in mobility patterns and the onset of symptoms of dementia.' With the aid of motion sensors placed throughout the person's home, both the total daily distance covered and average walking speed could be determined. When done over several years, changes in mobility patterns were able to predict the early stages of dementia.

- Two further health apps which are being trialled in 2017 are set to be introduced in the UK in 2018 by the NHS [72]. Both will enable patients to monitor their condition at home. For pregnant women with gestational diabetes, the app allows women to measure and send their blood glucose levels to their diabetes clinician. For chronic obstructive pulmonary disease, with 1–1.5 million sufferers in the UK alone, the app allows daily measurement of heart rate and blood oxygen saturation. After 3 months of measurement, the app learns that person's normal range, and is then able to alert the relevant healthcare professional when readings fall outside the normal range. As with other medical uses of big data, both apps have reduced visits to hospitals in trials.
- In Ontario, Canada, researchers are working with hospitals to improve the treatment of premature babies [56]. 'The software captures and processes patient data in real time, tracking 16 different data streams, such as heart rate, respiration rate, blood pressure, and blood oxygen level, which together amount to 1260 data points per second.'
- Another possible use is to help the disabled—those blind, deaf or in wheelchairs—find their way in buildings, again using a smart phone [3]. The system used wireless sensor networks with cameras and sensor nodes throughout the test building to help those handicapped to navigate the building.
- In Durham County, North Carolina, data from tax returns, the census, and lead concentrations found from blood tests, were integrated. It was thus possible to prepare maps of the county showing the high-risk areas for exposure to lead and improve detection and treatment for lead-affected children. Such data integration can 'bridge the chasm that has traditionally divided population health from clinical medicine at the individual level' [98].
- Big data has potential for mental health monitoring and improvement as well. In a 2017 New Scientist article [6], the social media site Facebook was reported as planning to use pattern recognition algorithms that could indicate whether people were suicidal from their online posts. In its planned trial of the idea, Facebook will make it easier for users to contact suicide prevention help centres.
- Google in April 2017 begun a 4-year trial collecting health- and lifestyle-related data from 10,000 volunteers [7]. It is hoped that by combining disparate kinds of data—from genes to physical activity, clues can be obtained for predicting the onset of cancer and other diseases.

- We have already discussed how the smart city will ideally have a multitude of fixed sensors throughout the city to continuously measure temperature, levels of various pollutants, pollen counts, etc. People with allergies could thus be advised on the best route to travel (and the best time) to avoid adverse health reactions [81]. Similarly, joggers could be advised on the best (i.e. healthiest) route for their run. The Sense2Health application combines both features of the Quantified Self (see Sect. 7.2.2) and real-time environmental condition monitoring. Users can therefore 'track, analyze and correlate well-being states with personal exposure levels to an environmental phenomenon, e.g., noise and do so with the least active involvement possible through automatic sensing (and bio-sensing)' [40].

7.2.2   Taking Charge: The Quantified Self Movement and Online Self-Help Groups

What several of these examples show is that there is already a recognisable trend in OECD countries toward future monitoring in the home of personal health status for people with various ailments, particularly the elderly. This trend is partly driven by the need to cut health costs in an ageing society. But there is already a nascent movement of people, termed the Quantified Self (QS) movement [69, 93], whose proponents measure and monitor their personal health signs, whether they are ill or not. Deborah Lupton [54] has reported that there are now already thousands of health-related applications available for smart phones.

According to Melanie Swan [84], the QS consists of 'individuals engaged in the self-tracking of any kind of biological, physical, behavioral, or environmental information as $n = 1$ individuals or in groups.' The idea of self-tracking is not new: in the eighteenth century, Benjamin Franklin monitored both his productivity and his moral character—the later based on a list of 13 virtues [31]. Nor is self-tracking today limited to a few individuals, as today '60% of U.S. adults are currently tracking their weight, diet, or exercise routine, and 33% are monitoring other factors such as blood sugar, blood pressure, headaches, or sleep patterns [84].' While tracking one's weight on a daily basis can be simply done and generates only one data point per day, the same is not true for heart rate monitors, which take about 250 readings every second in order to assess the risk of cardiac problems. Such a sampling rate would generate about 9 gigabytes every month for each person—clearly a big data problem if used for analysis [84].

As an example of the potential benefits of personal health tracking, an article in *New Scientist* [50] has reported how the use of body sensors can even alert the user to a potential health problem—in this case from Lyme disease. If vital signs such as heart rate and skin temperature are being continuously tracked by the individual, then changes in readings may give a better indication of a health problem than comparing a one-off reading with the average of the general population, as typically happens with a visit to a doctor.

Another example comes from people who record their running distances and speeds, and share this data with other runners on social network sites. Sinan Aral and Christos Nicolaides [8] have shown how personal data sharing can motivate members of such networks to run faster and further. It is possible that sharing other personal health- and well-being-related data with other interested persons could motivate those in the network for healthier lifestyles in general.

People are taking charge and assuming more responsibility for their own health in other ways. Persons suffering from various ailments, whether physical or mental, are already using social media sites for support and information, which may itself improve health outcomes [43]. One early study [38] looked at how diabetes sufferers—who are rapidly growing in numbers worldwide [99]—interacted on Facebook. The authors found that sufferers with this chronic disease used the site for various reasons: for mutual support, for finding (and sometimes sending) out more information about their illness and its treatment, and to gain more public recognition as a community. The researchers also found that various other actors were involved in these sites: family and friends of the patients, advertisers with relevant products to sell, and the medical research community (who may be interested in recruiting subjects for research participation). All these groups may have different motivations and levels of medical knowledge, and it may prove difficult to check whether the testimonials for products advertised are fictitious or not, or indeed whether or not those posing as researchers are really marketers.

The shift to more patient empowerment, while probably inevitable if future urban health problems are to be tackled, will not be without its own problems. Both chronic disease sufferers and healthy individuals will not only be sources of data for health researchers and practitioners but will often develop their own views on management of their disease or what is needed for healthy living, including diet. Inevitably much of the information their views are based on will be variously drawn from other QS enthusiasts, from social networks for specialist diseases like diabetes, and from general online health information websites. Even a casual perusal of these sites demonstrates that the quality of this information is very variable, varying as it does from information provided by renowned hospitals to the contradictory and thus confusing advice offered by myriads of alternative health websites [94]. Reasons for the popularity of alternative medicine treatments include low levels of scientific understanding, lower cost, and avoidance of side effects from many conventional therapies. Still, such alternative treatments long predated the Internet, and in any case are used by a minority of the population, particularly for serious illnesses.

7.2.3   Discussion

Melanie Swan [83] has also sketched an optimistic view of 'Health 2050', which differs in several ways from today's conception. First, a shift from a focus on management and cure of diseases toward an emphasis on prevention and well-being, a move which is already underway. A full shift will require attitude changes in both the general public and health care personnel. Despite the problems surrounding the future of QS should it become the norm, which have been well-articulated by Bietz et al. [12], a shift to QS or similar seems inevitable if health costs are to be cut and prevention and health monitoring replace cure as the dominant health paradigm.

7.3   Big Data Applications in Non-OECD Healthcare and Well-Being

Many of the nascent applications discussed in the previous section could be applied outside the OECD, given the ever-rising share of the global population who have access to smartphones and the Internet. Nir Kshetri [51, 52] is optimistic about the help big data can provide for industrialising economies in various sectors, including health. Local entrepreneurs are already providing big data services in the more technologically sophisticated and rapidly industrialising large economies such Brazil and India, but opportunities can also be found in smaller low-income countries.

Rosemary Wyber and her colleagues [98] have reviewed both the potential risks and the benefits from applying big data in low- and middle-income country health care systems. The list of the perceived risks ('dystopian views') are similar to those that face OECD countries but could be even more severe in low-income countries. They included the usual privacy issues of poor data governance and the effectiveness and meaning of patient consent; technical problems of data integration; and perhaps most important in the short term, diversion of limited resources, both financial and human, away from more effective public health approaches. When it is considered that many of the population will be illiterate and marginalised, or from minority groups, and that standards of governance are often very low, the risks from health and other personal information being misused could be high. As with the risks, the perceived general benefits ('utopian views') are similar to the promise of big data in industrialising countries, but could even lead to a 'major and beneficial turning point in global health' [98]. Included in the list are effective privacy rules that still allow the sharing of health data, and lower costs.

Section 7.3.1 briefly discusses actual applications of big data in industrialising economies. The actual role of big data in the Ebola outbreak, and how its usefulness could have potentially been enhanced is treated in Sect. 7.3.2, along with big data's potential for detecting other disease outbreaks. A final sub-section discusses the possible future of big data in healthcare in industrialising economies, along with the supporting policies necessary if the urban public are truly to benefit.

7.3.1   Examples of Existing Big Data Applications in Non-OECD Healthcare

- In India, what could be the beginnings of an ambitious national e-health scheme is already in place. The government has begun issuing 'Aadhaar' cards with a unique 12-digit identifying number to its vast population. The planning authorities are fully aware of the potential security and privacy risks and are taking steps to avoid them [82]. The potential benefits of this scheme when implemented derive from the possibility of both generating and monitoring health data and personal data on a huge scale—and using it to greatly improve public health [98].

- A better understanding of the industrialising world's slum populations and their growth, which can aid in providing crucial infrastructure—clean water, refuse collection, sanitation—is urgently needed. In the slum areas of Nairobi, Kenya's capital city, mobile phone data was used to provide the timely data needed [51]. Conventional sample surveys would be far more expensive, and, as discussed in Sect. 3.1, their findings will often be limited value if change is rapid.

- In Nigeria, population census results are out of date and often unreliable. This created problems with distributing of polio vaccines: some settlements were given too little, others too much. With funding from the Bill and Melinda Gates Foundation, satellite images and machine learning are combined to calculate population distribution. On the basis of the data generated, researchers can identify wealthier areas from their orderly street patterns and informal settlements by their denser housing and more random street patterns. Household population surveys can then be used to determine population densities for each area type [87].

- In Botswana, a pilot program—the Malaria Surveillance & Mapping project—begun in 2011. 'Health care workers are equipped with mobile phones to gather and upload malaria-related data to the cloud' [51]. They can also provide GPS data and images to Health Ministry officials.

7.3.2 The Role of Big Data in the 2014 West African Ebola Outbreak

Another potential application is for surveillance of outbreaks for diseases such as Ebola. Ebola is a very contagious disease with a high mortality rate, but previous outbreaks were fortunately of limited extent. The 2014 outbreak mainly impacted three low-income West African countries: Guinea, Liberia and Sierra Leone. These countries have few resources for both detecting and responding to such outbreaks. For example, Sierra Leone has only two doctors per 100,000 population, compared with 245 in the US [4]. Milinovich et al. [58] have argued that digital surveillance could improve response to outbreaks of this disease, as even in the low-income countries of west Africa, ownership of mobile phones is high and growing.

Fung and colleagues [37] have also pointed out another use made with big data approaches to dealing with Ebola disease. The US Centers for Disease Control (CDC) developed a model which can map projections of Ebola cases, with and without public health intervention. These easily understood maps proved a powerful means of *communicating* to relevant policy makers of the seriousness of the outbreaks and the consequences of inaction.

Twitter traffic, on-line news stories, and Internet searches about the disease could help pinpoint outbreaks in the early stages. Although this method may not be perfect, it may prove invaluable in low-income regions where more conventional disease surveillance methods are not well developed. Such use of social media and news feeds has already proved its worth for public health elsewhere. In 2002, the emergence of Sudden Acute Respiratory Syndrome (SARS) was first detected by a Canadian health news aggregator service.

7.3.3 Discussion

Many low-income countries lack both the necessary laboratory equipment and trained personnel, so getting tests analysed can be time-consuming. Smart phones potentially offer a way around this problem, and many of the health applications being developed could have even more relevance for such countries than they do for OECD countries, particularly as cellular networks there tend to be more reliable than both the electricity supply and even the availability of clean water. The apps being developed include one to detect schistosomiasis, a debilitating disease caused by parasitic infection. The test uses a mobile phone attachment to identify the presence of the parasites' eggs in urine and stool samples. Another app is a simple test for river blindness, another parasitic infection prevalent in Africa [66]. Such easily-obtained health care data could empower local people, and form the basis for surveillance for disease outbreaks.

Discussion of lifestyle changes—getting more exercise, eating healthier food and so on—is simply irrelevant for many residents of low-income cities, especially the hundreds of millions living in slums. They may have little choice in their lifestyle; they may not be able to avoid either drinking unsafe water or living in insanitary conditions. Unless these basic human needs are met, big data health applications can only be limited value in improving slum residents' health, even assuming the relevant technical expertise and infrastructure was available to health officials. Of course, better knowledge of, for example, the dynamics of informal settlements and the many problems facing their residents, could be helpful: the surveys of the poor in London and other cities by Seebohm Rowntree and others at the turn of the twentieth century were essential for improving their condition [96]. Even so, governments must be able and willing to provide the necessary resources to overcome these problems.

Nevertheless, governments in low-income countries do not necessarily have to rely solely on their own financial resources and technical expertise. In the case of Ebola, the risks were so great that international aid and expertise (from the WHO and especially Médecins Sans Frontières) complemented efforts by local health officials. More accurate and more detailed population estimates for settlements in Africa—and in fact anywhere in the world—benefitted from decades of high-resolution satellite images. Especially in the latter case, there are few problems with privacy.

7.4   Case Study: Instrumented Chair for Health and Comfort[1]

7.4.1   Introduction

An example of the Quantified Self is given here: the instrumented chair (the 'Virtual Spine' system) developed by the first author (in collaboration with Applied Health and IT researchers at Monash University Australia) to improve sitting posture and avoid back pain [90–92]. The application, like QS itself, is presently more relevant to OECD countries.

Sitting is one of the most common behaviors in people's daily life. A recent epidemiological study on about 50,000 adults from 20 countries reported sitting time was 300 min/day on average [11]. Incorrect sedentary positions and prolonged sitting has become a serious health threat to people in modern societies and results in various spinal problems [95]. A large number of studies have convincingly reported the association between different levels of exposure to occupational sitting and the presence or severity of low back pain (LBP) [74]. There is also unequivocal evidence that sitting and upper quadrant musculoskeletal pain are related [17]. The third European Working Conditions Survey identified the most common work-related health problem as back ache, reported by 33% of respondents [65]. These problems often result in absence from work or even permanent disability and translate into high economic costs [88].

It is a challenge for people to maintain appropriate sitting positions in daily life to avoid seating-related health issues. In the literature, discomfort and pressure sores have received particular attention in military [25], workplace [41] assisted living [59, 78] and mobility [1, 15, 80] contexts. For instance, the findings from Caneiro et al. [21] have demonstrated a clear link between thoraco-lumbar postures while sitting, and head/neck posture and motor activity. Burnett et al. [18] articulated the challenge to maintain appropriate sedentary behavior to reduce extra the burden to the spine and trunk muscles.

What is good posture? According to Claus et al. [24]: 'Good posture may be influenced by demands to prevent movement, coordinate movement, safely load spinal segments or conserve energy.' Despite the controversy around what constitutes an ideal sitting posture [64], it is clear that fixed postures, particularly in prolonged sitting, constitute a high-risk factor for developing LBP due to the static loading of soft tissues and discomfort [67]. As a preventive strategy, fixed postures in prolonged sitting should be avoided. Therefore, monitoring the spinal movements to prompt appropriate proactive measures is urgently needed.

Motivated by these findings, researchers from different disciplines are working on automatic monitoring of sitting postures to promote healthy sitting behavior. Specifically, clinicians are increasingly adopting spinal motion analysis as a useful

---

[1] This section is a slightly revised version of sections I–III of Wang and Yu (2013), Ref [92]. Used by permission of IEEE.

clinical method to quantify the range of trunk motion and pattern of posture changes for diagnosis and outcome evaluation. For example, skin surface tracking (from markers/sensors adhering to the skin overlying spinous processes) has been used to quantify spinal curve and the change in the lumbar spinal curve between positions from flexion to extension to identify optimized sitting posture [24]. Haller et al. [42] equipped an ergonomic office chair with four sensors to measure an office worker's posture and to produce different alerts to help people improving their sitting posture. However, the possibility of using a Pervasive Environment Simulator (PES) to collect and present through interactive 3D display detailed postural information has not been previously investigated.

Today, with the development of intelligent embedded agents and pervasive computing environments, performing experiments in a PES based game engine environment has become a cost-effective method to simulate users' real-time behavior [28]. The ideal model of a PES user monitoring and advisory system would comprise four seamlessly integrated modules:

- multiple sensors
- comprehensive data analysis (agents)
- interactive 3D (Mixed Reality)
- messaging.

Such a system enables multiple users to simultaneously view, discuss, and interact with the virtual 3D models, and enhance practice by supporting remote and co-located activities [13]. The multiple sensors module monitors various data about the users. A comprehensive agent model is designed to constantly monitor users' everyday activities [19]. Interactive 3D (Augmented Reality or MR [10]) displays provides great flexibility of viewpoint and intuitive interfaces to present information and support users to change their behavior [34]. Such a messaging system may act as powerful persuader because it can intervene in the right context as a convenient way to prompt users to change their behavior [14, 33, 100].

Ubiquitous computing and context-aware persuasive technologies [33] offer an opportunity to promote healthy behavior by presenting 'just-in-time', 'appropriate time' and 'appropriate place' information [45, 46]. An interesting discovery from this research is that the appropriate sedentary position varies depending on the *purposes* for which people sit: for example sitting in the office, in the car, or when dining, etc. The main focus of the Virtual Spine system design is to:

- monitor people's sedentary behaviour across various circumstances and encourage people to maintain appropriate sedentary positions under various contexts
- provide location-aware advice, for which the system relies on the 'chair id' to reflect the surrounding environment
- provide advice at the right moment, for which the system requires knowledge of the users' activities.

Advice on correct sedentary posture must fit easily into users' daily routine since messages suggesting simple activities are preferred over ones requiring significant effort [55]. Besides, lifestyle interventions can yield positive and long-term effects, in terms of increasing levels of moderately intense physical activity [36]. The suggested locations of lifestyle activities we include in our system are [36]: everyday activity (shops, homes, schools, workplaces, etc.) and recreation destinations (playgrounds, parks and gardens, etc.) Based on the current research and technical capabilities outlined above, in the next sections, we present a 'Virtual-spine' system that provides users with personalized and contextualized advice on appropriate sitting positions. We introduce the design, implementation and a planned evaluation test of the Virtual-spine system.

7.4.2   Implementation of the Virtual Spine

The Virtual-spine is implemented using the Unity3D and Arduino platforms. The Unity3D engine supports exporting application to mobile platforms, which will render the 3D images in real-time, based on the data from the Sedentary Position Analysis Unit.

The Virtual Spine consists of the following components:

- Sedentary Sensory Unit
- Advisory Unit
- Interactive 3D Unit
- Messaging Unit

Sedentary Sensory Unit

The Sedentary-Sensory Unit is designed to monitor users' spine movements while sitting by detecting the center of gravity and back curvature. This unit can be designed in suitable forms, such as pad-like shape, and set up in various places to gather sedentary position information in different contexts like at the office, at home or driving. Each unit has a unique ID used to recognize which chair the user is sitting on. This information is sent to the Advisory Unit which processes it to compute the cumulative spinal burden. Comprehensive sedentary information is presented to the user on mobile devices, computer screens or smart TVs as real-time rendered interactive 3D images. To reduce the burden on the spine, and to motivate users to maintain healthy sedentary habits, messages are presented pervasively using the most appropriate media. For instance, advice will be displayed on TV when the user is sitting on the sofa, or on a mobile device if the user is sitting on an office chair. Presentation settings could be tailored to users' privacy and other needs. The system architecture is illustrated in Fig. 7.1.

The Sedentary Sensory Unit provides high-sensitivity 'Pressure Sensing', 'Chair-ID Recognition' and 'Profile' functions. When users sit or even semi-sit on the sensing pad, their body affects the pressure sensors on it. Each sensor sends the detected values as analog input, which affects the 3D visualization through the Advisory Unit. Multiple physical pads should be placed on the seats habitually used for the different purposes. The pads' IDs determine the use context of each seat to provide correct posture parameters to the Sedentary Position Analysis unit. If the

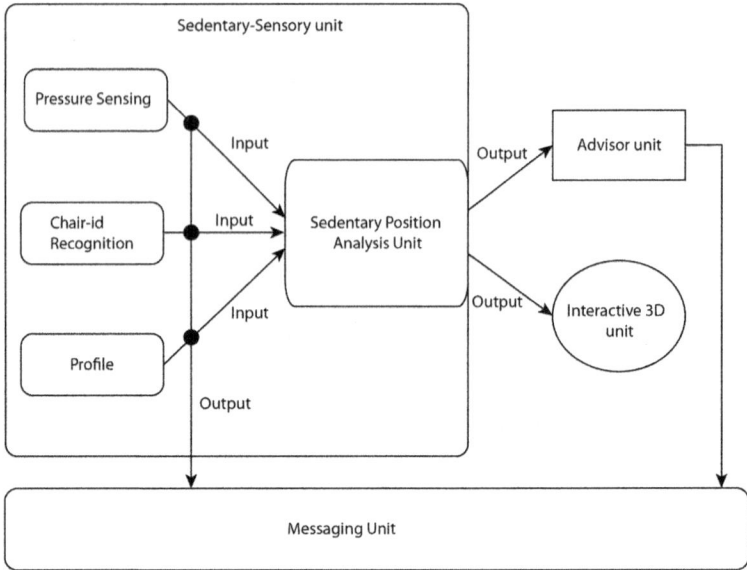

Fig. 7.1 System architecture for Virtual Spine

incorrect sedentary posture lasts over a set time threshold, the Advisor Service sends a message to the mobile UI.

1. *Pressure Sensing.* We used Force-Sensitive Resistor (FSR) sensors and an Arduino platform to detect postural changes. Sensors are laid out using a 'near-optimal sensor placement strategy' [61]. Any curvature variation of the body trunk is converted to analog input that affects the virtual 3D spine in real-time. The connection between Arduino and Unity3D is provided by a 'unity-Arduino serial connection' [68]. Figure 7.2 shows the Sedentary Sensory Unit (left), sensors affecting the virtual mat (middle), and interactive 3D spine (right).

2. *Chair ID Recognition.* Functionality and ergonomic characteristics such as seat height, depth, back support angle, surface material, flexibility (adjustment), etc. for each chair are stored in a database with a unique ID. The 'chair-ID' is recognized when users sit and *then* combined with users' profiles to calculate the most appropriate postures and sitting periods. The Chair-ID recognition function is based on Internet of Things technology. Currently, there are only a few companies providing free services to store IDs, profiles and input/output data in real-time (pachube.com, open.sen.se, etc.) Chair specifications could either be directly input or acquired from the manufacturers' database. The latter option also creates opportunities to design future 'smart' chairs, which could monitor ergonomic adjustments and send them to the manufacturers' database to inform design refinements. Users fill in their profiles through the online interface, providing information such as body size, gender, age, common sitting time, type of transportation, etc.

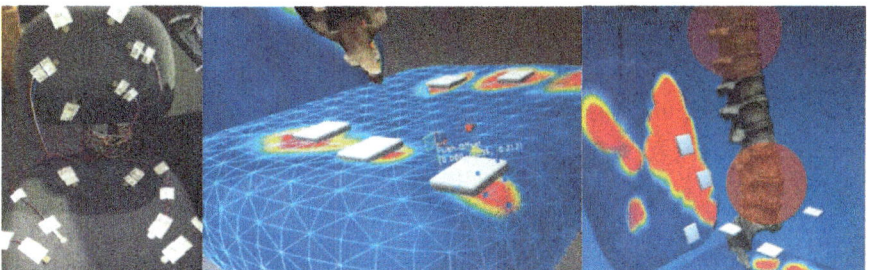

Fig. 7.2 Sedentary sensory unit

3. *Sedentary Position Analysis Unit.* Unhealthy postures are recognized based on duration and detected deviations from the ideal spinal position. Claus et al. [24] have suggested that the 'ideal' sitting position depends on the angles of three divided spine sections: 'thoracic, thoraco-lumbar and lumbar'. The posture detection system uses a comparison algorithm to analyze the spinal positions of each section. The chair-seat sensors are divided into four sections: a (left-front), b (left-back), c (right-front) and d (right-back) to work out the position of the lumbar area. The chair-back is divided into two sections: e (left) and f (right) to calculate the position of the thoracic area. The Analysis Unit compares the pressure input from these sections to calculate the thoraco-lumbar spine section position and movement. For instance, if the value of 'a' is much greater than 'b', 'c' and 'd' and both 'e' and 'f' are 0, then the user is heavily leaning to the left-front direction. However, if the values of 'e' and 'f' are also large at the same time, then the user is in a left leaning 'sloppy' position, as the thoracic area is positioned backwards.

Advisor Unit

Based on the posture recognition, chair-ID, and profile, the Advisor Unit calculates a suitability score for each advice in the activity database. Advice is generated through an expert system that considers the following factors:

- Spine angle: sharp angles cause extremely heavy burden on the spine and should not be maintained for a prolonged time
- Prolonged time: this parameter measures the duration of a position
- Accumulated sitting time: cumulative sitting time on different chairs to calculate total sitting time
- Frequency: how often the user takes the same position.

Instead of monitoring how the user follows the suggested activities, this system follows a decide-choose-do format to accumulate the chosen activities into the database. The Advisor Unit sends a query to one or more of the services and acquires their analysis results. If all the factors are met, the advice becomes a candidate. The advice mainly contains several types of message:

- Warning: to alert the user that it is time to change posture and stop sitting
- Activity: to suggest users take a proactive relaxation approach and what kind of activity is appropriate for their context. This type of message normally gives user several options to choose from depending on the contextual restrictions and time limitations
- Relaxation: to give the option to rest rather than doing exercises; this type of message suggests a minimum time-span during which any sitting should be avoided.

Here we present a sample scenario to show how it might work. Mr. A has been continuously working in his office chair for more than 3 h, and only maintained a healthy posture for less than 20% of this period. He will receive the following message: 'Please stand up straight with your arms at your sides, bend sideways to the left, slide your left hand down your thigh and reach with your right arm over your head. Hold this position for 10 s, then return to the starting position and repeat for the opposite side. Alternate sides for nine more times'.

Interactive 3D Unit

The 3D display unit presents spine information in two modes: a 'real-time mode' and an 'accumulation mode'. The real-time mode presents the user's current spinal position and corresponding burden. The accumulation mode presents the cumulative spinal burden information gathered during a certain period. Advice is mainly generated based on the accumulation mode data. It is very rare that a warning advice is directly generated from the real-time mode as this only happens when a user's movement results in an extreme burden to the spine.

Messaging Unit

Unlike conventional location-based pervasive functions, the Messaging Unit application uses the chair ID to recognise users' locations. This unit interacts with users by sending messages generated by the Advisory unit through the most appropriate medium.

7.4.3 Discussion

In the Virtual Spine lab unit, new techniques and materials are required to enable specifically purposed health monitoring, in this case, to improve spinal alignment by placing 'e-Skin' sensors on the spine area to monitor the key inter-vertebral disc

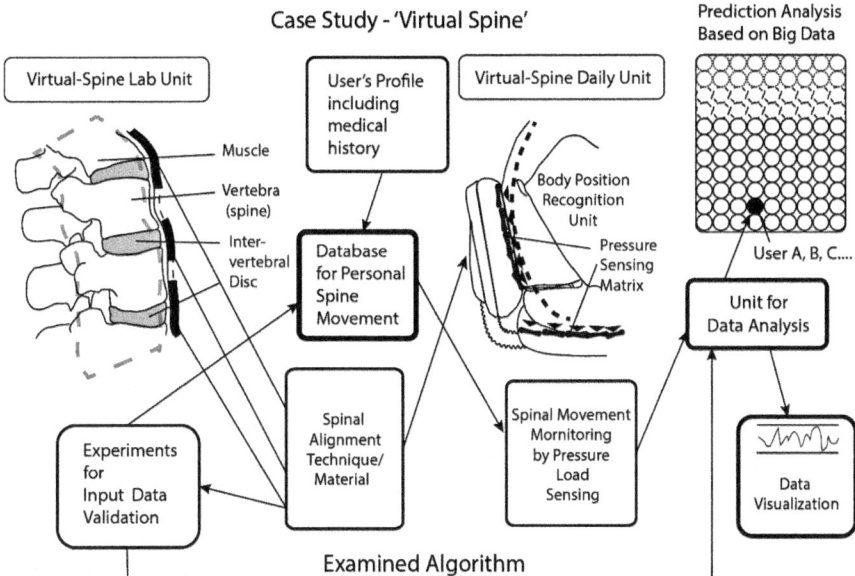

Fig. 7.3 The Virtual Spine platform as a case study for big data personal health support

expansions. The monitoring method will need to be validated through examining data examining data gathered from a series of tailored experiments. A database will then be built to store the samples of personal spine movement; this step is necessary to test, review and construct the appropriate database architecture. To provide personalized health support, a user's profile also needs to be built. The profile information will be used together with the spine movement data; both need to be contained in the database.

The Virtual Spine Daily Unit will need to be well designed, practical and attractive enough to be sold to various users to be part of people's daily life. The unit will be equipped with a pressure load sensing unit to enable spinal movement monitoring. This daily output data will be raw data which won't make any sense to users. However, based on the needs of users (e.g. medical practitioners and patients) some valuable/useful information (or knowledge) can be processed out with the unique algorithm from the outcomes from the validation experiments; this process will be achieved by the data analysis unit. According to various needs, the analyzed data can be visualised differently through a data visualisation unit.

The data, when analysed for all users, will be a big data set which could contribute to information on spine movement, spine health, sitting behavior, working habits, working condition for white-collar workers and so on. Figure 7.3 summarises how the data would be collected, and analysed.

7.5 Discussion: The Potential Benefits and Risks of Health Big Data

The use of big data for public health promises to bring many benefits but also carries some risks. As Vayena et al. [89] have pointed out, some of these downsides are present in public health regardless of the techniques used, while others are unique to big data approaches.

One possible danger with the potential for one version of personalised treatment promised by success in mapping the human genome is that notion of *public* health will be downgraded, and only the well-off will be able to afford medical care [97]. This is not only an equity problem. In most cities of the industrialising countries, contagious diseases are still a major cause of mortality, and ongoing climate change could increase the risk from these diseases. In the case of disease outbreaks like cholera or Ebola, the health of others affects the health of each individual.

Nevertheless, there is great potential for big data to improve health for all, both in the OECD and other countries, as illustrated in Sects. 7.2.1, 7.3.1 and 7.3.2. The QS movement—where people take more responsibility for their health—is a welcome development. The case study on the Virtual Spine, discussed in detail above, is an example of QS, and also suggests how the measurements for different individuals could be pooled to advance our knowledge.

References

1. Agency for Health Care Policy & Research (AHCPR) (1992) Pressure ulcers in adults: prediction and prevention, clinical practice guideline. AHCPR Publication no. 92-0047
2. Alsaqer M, Hilton B (2015) Indirect wayfinding navigation system for the elderly. Twenty-first Americas conference on information systems, Puerto Rico, 13pp. ://pdfs.semanticscholar.org/c5ef/6134b1956dcfe1dc45f4685c808dd63571df.pdf
3. Aly WHF (2014) MND$_{WSN}$ for helping people with different disabilities. Int J Distrib Sens Netw 10:7pp. https://doi.org/10.1155/2014/489289
4. Amankwah-Amoah J (2016) Emerging economies, emerging challenges: mobilising and capturing value from big data. Technol Forecast Soc Change 110:167–174
5. Anon (2016) Daunting data. Nature 539:467–468
6. Anon (2017) AI takes on suicide. New Scientist 233:6
7. Anon (2017) Big health data. New Scientist 234:4–5
8. Aral S, Nicolaides C (2017) Exercise contagion in a global social network. Nat Commun 8:14753. https://doi.org/10.1038/ncomms14753
9. Ausiello D, Lipnick S (2015) Real-time assessment of wellness and disease in daily life. Big Data 3(3):203–208
10. Azuma R et al (2001) Recent advances in augmented reality. IEEE Comput Graph Appl 21(6):34–47
11. Bauman A et al (2011) The descriptive epidemiology of sitting: a 20-country comparison using the international physical activity questionnaire (IPAQ). Am J Prev Med 41(2):228–235
12. Bietz MJ, Hayes GR, Morris ME, Patterson H, Stark L (2016) Creating meaning in a world of quantified selves. IEEE Pervasive Comput 15:82–85

13. Billinghurst M, Kato H (1999) Collaborative mixed reality. In: Proceedings of the international symposium on mixed reality (ISMR 99). Springer, Secaucus, NJ
14. Boland P (2007) Managing chronic disease through mobile persuasion. In: Fogg BJ, Eckles D (eds) Mobile perusasion: 20 perspective on the future of behavior change. Stanford Captology Media, Stanford, CA, pp 39–45
15. Brandeis GH et al (1994) A longitudinal study of risk factors associated with the formation of pressure ulcers in nursing homes. J Am Geriatr Soc 42(4):388–393
16. Brennan N, Oelschlaeger A, Cox C, Tavenner M (2014) Leveraging the big-data revolution: CMS is expanding capabilities to spur health system transformation. Health Aff 33(7):1195–1202
17. Brink Y, Louw QA (2013) A systematic review of the relationship between sitting and upper quadrant musculoskeletal pain in children and adolescents. Man Ther 18(4):281–288
18. Burnett A et al (2004) Spinal kinematics and trunk muscle activity in cyclists: a comparison between healthy controls and non-specific chronic low back pain subjects. Man Ther 9:211–219
19. Callaghan V et al (2004) Intelligent inhabited environments. BT Technol J 22(3):233–247
20. Campo E, Grangereau E (2008) Wireless fall sensor with GPS location for monitoring the elderly. 30th annual international IEEE EMBS conference, Vancouver, BC, Canada, August 20–24, pp 498–501
21. Caneiro J et al (2010) The influence of different sitting postures on head/neck posture and muscle activity. Man Ther 50(1):54–60
22. Chawla NV, Davis DA (2015) Bringing big data to personalized healthcare: a patient-centered framework. J Gen Intern Med 28(Suppl 3):S660–S665
23. Clark J (2014) Do the solutions for global health lie in healthcare? Br Med J. https://doi.org/10.1136/bmj.g5457
24. Claus AP et al (2009) Is 'ideal' sitting posture real?: measurement of spinal curves in four sitting postures. Man Ther 14(4):404–408
25. Cohen D (1998) An objective measure of seat comfort. Aviat Space Environ Med 69(4):410–414
26. Cook D (2012) How smart is your home? Science 335:1579–1581
27. Craven M, Page CD (2015) Big data in healthcare: opportunities and challenges. Big Data 3(4):209–210
28. Davies M, Callaghan V, Shen L (2007) Modelling pervasive environments using bespoke and commercial game-based simulators. Lect Notes Comput Sci 4689:67–77
29. De la Torre Díez I, Cosgaya HM, Garcia-Zapirain B, López-Coronado M (2016) Big data in health: a literature review from the year 2005. J Med Syst 40:209. https://doi.org/10.1007/s10916-016-0565-7
30. Dye C (2008) Health and urban living. Science 319:766–769
31. Farrington C (2016) Big data meets human health. Science 353:227
32. Fleming E, Haines A, Golding B et al (2014) Data mashups: potential contribution to decision support on climate change and health. Int J Environ Res Public Health 11:1725–1746. https://doi.org/10.3390/ijerph110201725
33. Fogg BJ (2003) Persuasive technology: using computers to change what we think and do. Morgan Kaufmann, San Francisco, CA
34. Fogg BJ, Eckles D (eds) (2007) Mobile persuasion: 20 perspectives on the future of behavior change. Stanford Captology Media, Stanford, CA
35. Ford JD, Tilleard SE, Berrang-Ford L et al (2016) Big data has big potential for applications to climate change adaptation. Proc Natl Acad Sci U S A 113:10729–10732
36. Frank L, Engelke P, Schmid T (2003) Health and community design: the impact of the built environment on physical activity. Island Press, Washington, DC
37. Fung IC-H, Tse ZTH, Fu K-W (2015) Converting big data into public health. Science 347:620
38. Greene JA, Choudhry NK, Kilabuk E et al (2010) Online social networking by patients with diabetes: a qualitative evaluation of communication with Facebook. J Gen Intern Med 26(3):287–292

39. Groves P, Kayyali B, Knott D, Van Kuiken S (2013) The 'big data' revolution in healthcare. Available at http://www.pharmatalents.es/assets/files/Big_Data_Revolution.pdf
40. Hachem S, Mathioudakis G, Pathak A (2015) Sense2Health: a quantified self application for monitoring personal exposure to environmental pollution. SENSORNETS 2015. https://hal.inria.fr/hal-01102275/document. Accessed 30 Mar
41. Halender MG, Zhang L (1997) Field studies of comfort and discomfort in sitting. Ergonomics 40(9):895–915
42. Haller M et al (2011) Finding the right way for interrupting people improving their sitting posture. In INTERACT 2011, Part II, LNCS 6947, pp 1–17. © IFIP International Federation for Information Processing
43. Hobbs WR, Burke M, Christakis NA (2016) Online social integration is associated with reduced mortality risk. Proc Natl Acad Sci U S A 113:12980–12984
44. Hodson H (2016) Google knows your ills. New Scientist 230:22–23
45. Intille SS (2004) A new research challenge: persuasive technology to motivate healthy aging. IEEE Trans Inf Technol Biomed 8(3):235–237
46. Intille SS (2004) Ubiquitous computing technology for just-in-time motivation of behavior change. Stud Health Technol Inform 107(2):1434–1437
47. Ioannidis JP (2005) Why most published research findings are false. PLoS Med 2(8):e124. (PMID: 16060722)
48. Kellermann AL, Jones SS (2013) What it will take to achieve the as-yet-unfulfilled promises of health information technology. Health Aff 32(1):63–68
49. Khoury MJ, Ioannidis JPA (2014) Big data meets public health. Science 346:1054–1055
50. Klein A (2017) Alert: you're about to catch something. New Scientist 233:16
51. Kshetri N (2014) Big data's impact on privacy, security and consumer welfare. Telecommun Policy 38:1134–1145
52. Kshetri N (2016) Big data's big potential in developing economies: impact on agriculture, health and environmental security. CABI Press, Boston
53. Lazer D, Kennedy R, King G et al (2014) The parable of Google Flu: traps in big data analysis. Science 343:1203–1205
54. Lupton D (2013) Quantifying the body: monitoring and measuring health in the age of mHealth technologies. Crit Public Health 23(4):393–403
55. Maheshwari M, Chatterjee S, Drew D (2008) Exploring the persuasiveness of "Just-in-time" motivational messages for obesity management. In: PERSUASIVE '08 proceedings of the 3rd international conference on persuasive technology. Springer, Berlin
56. Mayer-Schönberger V, Cukier K (2014) Big data. Mariner Books, Boston
57. McGrath MJ, Ni Scanaill C (2015) Sensor technologies: healthcare, wellness and environmental applications. Apress Open, New York
58. Milinovich GJ, Soares Magalhães RJ, Hu W (2015) Role of big data in the early detection of Ebola and other emerging infectious diseases. Lancet Glob Health 3(1):e20–e21
59. Monette M, Weiss-Lambrou R, Dansereau J (1999) In search of a better understanding of wheelchair sitting comfort and discomfort. In RESNA annual conference
60. Moriarty P, Honnery D (2015) Future cities in a warming world. Futures 66:45–53
61. Mutlu B et al (2007) Robust, low cost, nonintrusive sensing and recognition of seated postures. In: UIST '07 proceedings of the 20th annual ACM symposium on user interface software and technology. New York, ACM
62. Neff G (2013) Why big data won't cure us. Big data 1(3):117–123
63. Obermeyer Z, Emanuel EJ (2016) Predicting the future—big data, machine learning, and clinical medicine. N Engl J Med 375(13):1216–1219
64. O'Sullivan K et al (2012) What do physiotherapists consider to be the best sitting spinal posture? Man Ther 17(5):432–437
65. Paoli P, Merllie D (2001) Third European survey on working conditions 2000. European Foundation for the Improvement of Living and Working Conditions, Luxembourg
66. Perkel JM (2017) Pocket laboratories. Nature 545:119–121
67. Pope M, Goh K, Magnusson M (2002) Spine ergonomics. Annu Rev Biomed Eng 4(1):49–68

68. Project: unity-arduino with serial connection. http://code.google.com/p/unity-arduino-serial-connection/. Accessed 27 Feb. 2013
69. Quantified Self (2016) Available at http://quantifiedself.com/guide/
70. Raghupathi W, Raghupathi V (2014) Big data analytics in healthcare: promise and potential. Health Inf Sci Syst 2(3):1–10
71. Revell T (2017) AI turns phone into Parkinson's test. New Scientist 233:14
72. Reynolds M (2017) NHS to prescribe apps that keep tabs on health. New Scientist 233:11
73. Reynolds M (2017) Smart meter knows if you need help. New Scientist 233:12
74. Roffey D et al (2010) Causal assessment of occupational sitting and low back pain: results of a systematic review. Spine J 10(3):252–261
75. Roski J, Bo-Linn GW, Andrews TA (2014) Creating value in health care through big data: opportunities and policy implications. Health Aff 33(7):1115–1122
76. Rutkin A (2016) Medicine by machine. New Scientist 231:20–21
77. Shah NH (2015) Using big data, Chapter 7. In: Payne PRO, Embi PJ (eds) Translational informatics: realizing the promise of knowledge-driven healthcare, Health informatics. Springer, London. https://doi.org/10.1007/978-1-4471-4646-9_7
78. Shaw G (1992) Wheelchair seat comfort for the institutionalized elderly. Assist Technol 3(1):11–23
79. Sheth A, Anantharam P, Henson C (2013) Physical-cyber-social computing: an early 21st century approach. IEEE Intell Syst 28:78–82
80. Smith DM (1995) Pressure ulcers in the nursing home. Ann Intern Med 123(6):433–438
81. Solanas A, Patsakis C, Conti M, Vlachos IS et al (2014) Smart health: a context-aware health paradigm within smart cities. IEEE Commun Mag 52:74–81
82. Srivastava S, Agarwal N, Agarwal R (2013) Authenticating Indian E-Health system through "Aadhaar" a unique identification. Int J Sci Eng Res 4(6):2412–2416
83. Swan M (2012) Health 2050: the realization of personalized medicine through crowdsourcing, the quantified self, and the participatory biocitizen. J Pers Med 2:93–118. https://doi.org/10.3390/jpm2030093
84. Swan M (2013) The quantified self: fundamental disruption in big data science and biological discovery. Big Data 1(2):85–98
85. Taleb N (2013) Beware the big errors of 'big data'. Wired blog. http://www.wired.com/2013/02/big-data-means-big-errors-people/. Accessed 15 Dec 2015
86. Tatem AJ, Hay SI, Rogers DJ (2006) Global traffic and disease vector dispersal. Proc Natl Acad Sci U S A 103:6242–6247
87. Tollefson J (2017) Satellite images reveal gaps in global population data. Nature 545(7653):141–142. https://www.nature.com/news/satellite-images-reveal-gaps-in-global-population-data-1.21957
88. van Tulder M, Koes B, Bouter L (1995) A cost-of-illness study of back pain in The Netherlands. Pain 62(2):233–240
89. Vayena E, Salathé M, Madoff LC et al (2015) Ethical challenges of big data in public health. PLoS Comput Biol 11(2):e1003904. https://doi.org/10.1371/journal.pcbi.1003904
90. Wang SJ (2014) Seated posture detection. Monash University, Australia. Provisional Patent, File No. 2014901381 (Patent Pending)
91. Wang SJ (2014) System and method for monitoring user posture. Monash University, Australia. International Patent Application No PCT/AU2014/000488 (Patent Pending)
92. Wang SJ, Yu D (2013) Virtual-spine: the collaboration between pervasive environment based simulator, game engine (mixed-reality) and pervasive messaging. In Proceedings of the 7th international conference on pervasive computing technologies for healthcare and workshops (ICST), pp 45–48. doi: https://doi.org/10.4108/icst.pervasivehealth.2013.252108, https://doi.org/10.4108/icst.pervasivehealth.2013.252108
93. Wikipedia (2016) Quantified self. Available at https://en.wikipedia.org/wiki/Quantified_Self
94. Wikipedia (2017) Alternative medicine. Available at https://en.wikipedia.org/wiki/Alternative_medicine
95. Wikipedia (2017) Back pain. Available at https://en.wikipedia.org/wiki/Back_pain

96. Wikipedia (2017) Seebohm Rowntree. Available at https://en.wikipedia.org/wiki/Seebohm_Rowntree
97. Wilbanks JT, Topol EJ (2016) Stop the privatization of health data. Nature 535:345–348
98. Wyber R, Vaillancourt S, Perry W et al (2015) Big data in global health: improving health in low- and middle-income countries. Bull World Health Organ 93:203–208
99. Zimmet P, Alberti KGMM, Shaw J (2001) Global and societal implications of the diabetes epidemic. Nature 414:782–787
100. Boland P (2007) Managing chronic disease through mobile persuasion. In: Fogg BJ, Eckles D (eds) Mobile persuasion. 20 perspectives on the future of behavior change. Stanford Captology Media, Stanford, CA, pp 45–52

Chapter 8
Big Data for a Future World

8.1 Introduction: The World in 2050

What does the future hold? Although to an increasing extent we are constrained by the biophysical limits of a finite planet, we believe there is still the opportunity for all humanity to live a full life within these limits. But we need to learn the lessons natural ecosystems teach us: first, do not use up natural resources faster than the Earth system regenerates them, and second, limit waste production, including GHGs, to the rate at which they can be safely assimilated by the environment [10].

In this introductory section, we first attempt to picture the world in general in 2050, stressing the changes from the present-day. It is, in some ways at least, an optimistic view, but we outline arguments to support this position. In Sect. 8.2 we examine how the world in 2050 might actually use big data for urban sustainability in both OECD and non-OECD cities, and how we might get to this 2050 world from where we at present. Finally, Sect. 8.3 outlines the difficulties that must be overcome in the coming years if big data is to fulfil its potential for urban sustainability.

8.1.1 A Changing Planet

The year is 2050. The population of Earth, only about 1.65 billion at the turn of the twentieth century, is now some 9.725 billion. Globally 4.5% of these are aged 80 years or more, up from 1.7% in 2015, but in several countries, the proportion has already reached 15% [41]. Most of the world's population live in cities, even though in some countries a counter-movement back to smaller towns and rural areas has occurred [31]. It has now become evident to all that the climate and other elements of the Earth's environment have decisively changed, and that the world is entering unchartered territory. Atmospheric CO_2 levels are now at their highest level since 50 million years ago [20]. Because of the decisive changes that the fossil fuel era left even in the

© Springer International Publishing AG, part of Springer Nature 2018
S. J. Wang and P. Moriarty, *Big Data for Urban Sustainability*,
https://doi.org/10.1007/978-3-319-73610-5_8

geological record, a new geological epoch has been proposed—the *Anthropocene* [43]. Fossil fuel use is now heavily in decline, caused by both global depletion of readily accessible reserves, and heightened awareness of the external costs of their production and use, principally, but not only, their GHG emissions. Not only fossil fuels, but annual production of several essential minerals are also in decline. In the memorable phrase of Richard Heinberg [12], we are now in an era of 'peak everything'.

Global temperatures have continued to rise, although the increase is now slowing. Because of the long atmospheric lifetime of CO_2, concentrations continued to rise even after annual emission levels were cut. Nevertheless, for most people, this temperature rise is imperceptible, given the far larger daily and seasonal temperature variations people in temperate climates, at least, experience. What is more important, however, is the increased frequency and occurrence of extreme weather events, particularly prolonged heat waves, and intense rainfall, that is already underway and predicted to intensify [11, 13]. It is this change, increasingly personally experienced by most of the world's population, that has led to political pressure for urgent climate change mitigation. For many coastal cities and other coastal settlements, these concerns have been joined by sea level rise [19]. One important response (after much political wrangling) was the imposition of carbon quotas for each country, with such national quotas declining each year.

National and global GDP growth are no longer at the forefront of national and international policy. It had long been recognised that GDP was a poor measure of human welfare. Indeed, Ida Kubiszewski and her colleagues [18] had shown in the early 2010s that for many OECD countries, alternative measures of welfare such as the Genuine Progress Indicator showed a steady fall, even as GDP per capita continued to rise. (Nevertheless, in industrialising countries, both measures were still then showing steady increases.) This discrepancy in the OECD countries, along with concerns about Earth ecological limits, prompted some researchers to advocate *degrowth*, or a planned reduction in GDP. On the other hand, Jeroen van den Bergh [42] instead advocated what he called *agrowth*, or in other words, that we should stop worrying about conventional GDP trends, and concentrate on solving the world's social and ecological problems. This suggestion has been taken up, implicitly at least. One important outcome of this change is that in this ecologically constrained world, consumerism has lost much of its force. A product of the growth economy, it was simply out of step with the new realities.

It is now widely appreciated that in earlier decades, the people of Earth were living unsustainably, with the global ecological footprint already about 1.5 times the Earth's area in the late-2010s. In other words, the Earth was already in *ecological overshoot*, a condition which could only be temporary if we were to live sustainably [36]. As stressed by Will Steffen and his colleagues [38] in the early years of this century, the Earth was perilously close to a variety of ecological and resource limits. Because of these multiple limits, many proposed tech fixes for one limit were found to exacerbate one or more of the other limits, necessitating the measures discussed below [32]. For example, SRM might successfully reduce global temperatures, but at the expense of adverse changes in precipitation in some regions, and progressive ocean acidification would continue. Even fully-justified attempts to reduce national SO_2 emissions meant that global warming was increased, because SO_2 pollution increases local albedo, and so reduces ground level insolation.

Just as the successors to the earlier Quantified Self movement monitor a variety of personal health indicators, the health of the entire planet (including its cities) is now closely monitored—the condition of its soils, atmosphere, oceans, rivers and lakes, ice caps, agricultural lands and pastures, forests and ecosystems are sampled continuously by a vast array of sensors, to determine their change over space and time. These sensor outputs are vital in enabling us to formulate plans to keep within safe ecological boundaries.

This global- and local-scale sampling and data collection were partly made possible by a multitude of people who act as citizen scientists. This was not a new idea: data on bird and even insect numbers and their distribution have been provided by non-scientists for many decades. Today, large numbers of citizens from all over the world use their smart phones to provide a great variety of data. In the environmental and health sciences, at least, the role of these citizen scientists are seen as indispensable [23]. They are motivated by their concern for the environment and their desire to be part of the scientific endeavour. The sciences have regained their prestige among the general population, as they are regarded as vital to balancing the often conflicting, but also partly complementary, demands of humans and nature.

For many, this data collection and entry is even the main way in which they use smart phones. They measure local air and noise pollution levels, record sightings of birds and native animals, and provide pictures of everything from local urban watercourses and trees (from which their ecological health can be determined) to damaged urban footpaths. Local residents take readings of stream levels and flows in places not readily accessible, and monitor local crop conditions and food prices [5].

Despite continuing national rivalries, governments now realise the global interdependency of nations; they appreciate that we increasingly face a common future, that the multiple environmental and resource limits we face are *planetary* limits [38]. As an example, consider that air pollution generated in North-East China blew over the Yellow Sea and Sea of Japan to Korea and Japan, and even to the west coast of the US [22]. Fires in Indonesia created air pollution in Malaysia and Singapore and were hazardous for air travel in the region. The future of the Amazon rain forest, the Russian tundra, and the Greenland icecap is of continuing concern to all people, because of their global climate change implications. Because of this, enlightened self-interest has made the provision of real assistance to the lower income countries of the late 2010s far more politically acceptable to the people in OECD countries. Per capita living standards across countries now show far less variation than those shown for the year 2014 in Table 1.1, and income distribution within countries is now more even.

8.1.2   Responses to a Changing Planet

Beginning in several OECD countries, car use per capita peaked in the first decade of the current century, later followed by absolute declines in both car numbers and vehicle-km driven. Eventually, this pattern was repeated worldwide, although in countries outside the OECD, car ownership and use continued to rise for some time after peaking in the OECD. It was finally realised that continued car-based transport was incompatible with the strict national quotas for CO_2 and other GHG emissions necessary for climate stabilisation. It was further realised that vehicular travel—particularly in OECD countries—had grown to excessive levels in the era of cheap fossil fuels: globally, it had risen from a mere 0.20 trillion passenger-km (all by rail) in 1900 to about 44 trillion passenger-km by all modes, including air, in 2010, but peaked a decade or so later. Global air travel grew from almost zero in 1950 to 6.6 trillion passenger-km in 2015 [1, 28]. Although in 2016 it was originally forecast by the industry to double over the 2016–2035 period, air travel has also been cut back significantly, for reasons similar to those for surface travel reductions.

For cities, urban travel overall has also been greatly reduced, with non-motorised travel and public transport largely replacing private travel in this reduced total [27]. It was realised that many existing urban trips were under two km, and so feasible for walking and cycling trips; more importantly, many more distant intra-urban destinations preferred in an age of cheap fossil fuels have been replaced by closer ones. It is now recognised that much of this travel was ultimately unnecessary, a product of both cheap oil and planners who encouraged travel growth as a sign of urban progress.

In a manner analogous to transport, total commercial energy use has fallen significantly, even as global population has risen. An important reason for the initial decline was the removal of heavy subsidies for fossil fuels: much of the world's high energy use in the late 2010s was an artefact of these high subsidies. Renewable energy sources have replaced fossil fuels as the dominant energy source, with nuclear energy continuing to have a minor share. Subsidies for these sources now have also been removed. Since the global supply of geothermal, hydro and biomass energy is limited [33], wind and solar provide not only most electricity but most energy for non-electric uses as well. Both because these forms of energy are intermittent, and in the case of solar PV electricity, millions of households act as producers from rooftop installations, smart grids and real-time pricing of electricity are needed. To enable a better demand match to this variable supply, domestic electricity demand is extensively managed, in two ways. First, for continuously operating domestic—and non-domestic—appliances such as boilers, refrigerators, and freezers, short-term energy supply fluctuations are managed by automatically switching them off for short periods when intermittent energy supply is low. For longer-term supply fluctuations, demand is reduced by *activity shifting*. Based on supply forecasts, households (and businesses) are motivated by much lower unit costs to shift activities like clothes washing to periods of plentiful intermittent energy supply. With smart appliances, such shifting can be done automatically.

The limited availability of important minerals was also of rising concern by the year 2020 and beyond. Paradoxically, it was finally realised that cities themselves represented a source of many minerals in the form of waste buried in landfill dumps and even in the dust beside urban roads, and abandoned infrastructure such as underground pipes [37]. One difficulty facing recovery and recycling was the often low concentrations of the resource. Paradoxically, this was often the result of efficiency gains, such as using less platinum catalyst in three-way catalytic converters. The low quantities available made recycling less feasible. But legislation now mandates design for reuse, so that the full life cycle of products is now the basis of design, instead of the manufacturing phase only.

Serious policies at the global level for climate mitigation were late in getting underway, partly because of the high hopes placed on technologies that would have potentially avoided the need to reduce fossil fuel use. Indeed, atmospheric CO_2 levels continued to rise for decades after the IPCC published their first report in 1991. These proposed technology fixes included carbon capture and storage (CCS) and geoengineering. CCS, among other difficulties, proved to be too expensive in both energy and money terms and has been largely abandoned as a major solution. Geoengineering—given the very unequal geographical (and thus political) distribution of benefits and costs—proved too politically divisive, and in any case did nothing to alleviate progressive ocean acidification, and the major ecosystem changes—and hence risks—it entailed.

As well as intense mitigation efforts, *adaptation* to ongoing climate change has proved necessary. Adaptation to climate—and other environmental—changes has taken many forms. In urban areas, two important challenges have been the increased risk of flooding brought about by both increased rainfall intensity and rising sea levels, and the health effects of extreme heat waves. In many cases, flood protection barriers were seen to be not only very expensive but also merely to shift the problem to less protected areas. Even within cities, the unequal risks borne by different income groups became clear, particularly in low- and middle-income cities, where the poor often lived in unplanned settlements on flood plains, or steep slopes susceptible to landslides after heavy rains. In some cases, the coastal parts of cities, or even entire cities, have had to be abandoned, despite the heavy write-off in real estate values entailed [31].

The increased frequency and severity of heat waves in cities has been exacerbated by the UHI effect. The usual method of adaptation in middle- and high-income urban households, air conditioning, not only had high power demands but also exacerbated the UHI by its inevitable waste heat release. While unfortunately necessary, limits are now placed on household and office air conditioning use. This use reduction has been partly achieved by recognising the value of personal *acclimatisation* to raised temperature levels [3]. (Acclimatisation refers to the ability of humans to adjust to sustained warm temperatures over time. Moving in and out of air-conditioned buildings can interfere with this adjustment.) Clothes are now worn that are suited to temperature rather than solely for fashion, and corresponding adjustments have been made to office dress codes. All these methods are based on individual adjustment to higher temperatures. But another important means of coping is the use of *passive cooling* of buildings, an approach used for many centuries in the vernacular architecture of countries with hot climates. Although most effective (and cheapest) if incorporated as part of the original design, buildings can be retro-fitted to take advantage of passive cooling techniques. Even with passive cooling, the actions of occupants are critical to its implementation: occupants need to open windows for cross ventilation at appropriate times, operate sun shades on windows and minimise heat from cooking in hot weather, and even change rooms they are occupying during the day.

Healthcare has also been radically transformed, driven not only by demands for higher quality care but also by the urgent need to cut total healthcare costs in an ageing society. Economic growth is no longer a possible solution to rising expenditures. It is now well-recognised that the health of residents is only partly determined by conventional healthcare, and that diet, local pollution levels, amount of exercise, and stress levels all have a major influence on both physical and mental health. One important change is that the general public is expected to take a far more active role in their own health and well-being. The Quantified Self movement has spread from a small group to embrace the general population in one form or other. People are now routinely expected to monitor many of the body's vital signs from home, particularly those that suffer from chronic diseases such as diabetes. But many healthy people now routinely input nutritional data from the food they eat, log their exercise levels, weight, blood pressure levels, and so on. Initially, there was much confusion about the proper role of physician and patient; patients were more empowered and obviously had a vital interest in their own health, but the downside of this was often a conflict with the general medical expertise of the doctor, because of the heterogeneous quality of medical information available to the public from the Internet. Eventually, a new healthcare model emerged, with the roles of both patient and doctor different from the traditional ones, and more clearly defined.

All sections of the urban population have now finally come to appreciate that their own health, well-being, and quality of life are not independent of these same factors for their fellow urban residents. As an example, excessive use of antibiotics—much of it for livestock—led to the rise of anti-biotic resistance. Contagious diseases are another obvious example of this interdependence, and eventually, the strong link between vast income inequality and social instability also encouraged more equitable policies. The Quantified Self for individuals was seen as insufficient: the health of the city residents as a whole was also important, and numerous indicators of urban health have been developed to regularly track this important factor.

Overall, the main change is that both ordinary people and their policy makers now view the biophysical world in a manner similar to the minority perspective held by a handful of Earth system scientists earlier in the century. Foremost is the common acceptance of the notion of global limits and its inevitable political and economic implications. Beginning with GHG emissions, the nations of the world had to decide how to allocate the allowable CO_2 emissions between countries [8]. Should it be based on cumulative total historical emissions, or just on present emission levels? What about net exports or imports of embodied CO_2 emissions? Apart from allocating GHG emissions, similar questions arose for allocating supplies of oil and scarce minerals between nations. The first step was the removal of the vast subsidies for energy, especially fossil fuels, and the introduction of a significant, and annually increasing, price for carbon. Much of the revenue collected is being directed to poorer countries or individuals within countries—so that environmental sustainability complements measures to combat inequality. Import tariffs were imposed on countries that operated without carbon pricing to prevent free-loading [21]. The debate over the best methods of GHG allocation continue, and further changes are expected, but deep reductions in emissions are finally occurring.

The second, related, change was the widespread acceptance of the interconnectedness of topics and sectors formerly viewed as separate (and so treated by policy makers, at all levels—regional, national and international). Urban transport is related to urban health in several ways, as discussed earlier. Likewise, urban energy use cannot be separated from concerns about air and noise pollution, and the UHI effect. Urban health, both mental and physical, and general well being and life satisfaction depend on urban planning and the provision of green spaces and nature reserves in urban areas, and on the levels of income and social inequalities within the city. They likewise obviously depend on the provision of non-polluted water and an adequate diet.

Throughout the early history of the Internet, there had been a tension between idealistic views of its potential, as exemplified on the one hand both by Wikipedia and free software development, and by the rise of citizen science, as discussed above, and on the other hand, by its commercial applications. It was finally realised that the shady practices of many early internet businesses were not only alienating people but were eventually counter-productive, and so regulations were progressively tightened. As an example, Google announced that beginning in 2018, it would remove the most intrusive ads, with the hope that viewers would tolerate the remaining ads [2]. Also, with the global shift in emphasis from economic growth and consumerism to ecological sustainability, advertising, in general, is less important. The Internet as a source of information is once again to the fore. Social sites are still popular, but the people alive in 2050—who have never known a world without the internet, smart phones and their successors—are now much more circumspect about what they post online.

8.2   The Role of Big Data in Cities in 2050

The optimistic description of the world in 2050 discussed above raises several important questions. How do we get from where we are today to this future? What is the role of big data in this transition? Will there be common elements in the transition for all cities, or will it be very different for cities in different parts of the world? An attempt to answer these questions is made in the following sub-sections.

8.2.1   Big Data in OECD Cities

In Sect. 8.1 we have described in general terms what the world of 2050 might look like, as nations everywhere grapple with the increasingly evident multiple environmental constraints. In the present section, we discuss how big data applications can help bring about this sustainable world, starting with the views of three writers on this topic who were all writing from the viewpoint of the future in OECD countries. In 1997 computer guru Vinton Cerf [7], in an edited book written to celebrate the first half-century of computers, attempted a description of what the computerised home—what is now called a smart or intelligent home—would look like in 50 years time, i.e. in 2047. While the article is remarkably prescient, with many of his forecasts already coming to pass, what is lacking is any appreciation of global environmental or resource limits, or indeed the use of IT to further sustainability aims. Instead, the emphasis was on 'Robert', an American from New Mexico, and his conversations with the house computer ('Jeeves'), who has arranged his extensive schedule for that day, which included a lecture in New York, lunch in Los Alamos, and then back home to Taos. Evidently, the almost ten billion people expected to live on our planet in 2050 cannot live like 'Robert', particularly given his 'hypermobile' lifestyle.

In 2015, Bill Montgomery in a brief article outlined a possible future for big data in the year 2029 [25]. Like Cerf, he was optimistic about the benefits the Internet (in the form of widespread application of big data and the IoT) will bring to the world. He stressed that such applications would go well beyond its mundane uses for traffic optimisation and lighting energy minimisation. As early as 2029 he foresaw an energy-efficient global society in place, with domestic energy consumption reduced through the introduction of smart meters and smart appliances. For transport, he argued that widespread availability of fully automated vehicles would reduce the need for car ownership, resulting in less road traffic. IoT sensors in buildings and other infrastructure such as bridges would be used to give advanced warning of structural problems.

Montgomery also argued that IoT applications would also protect natural resources such as forests, with sensors to monitor soil moisture levels, and help mitigate the damage from forest fires. Similarly, in the agricultural sector, sensors would monitor soil moisture and temperature at all times, enabling greatly increased production of basic foodstuffs. For human health, wearable devices would track a variety of vital health indicators, providing information to both the individual and health workers, and communicating appropriate actions that need to be taken.

Paul Buddle [6] had a similarly optimistic view of the potential of big data and smart cities in his chapter entitled *Smart Cities of Tomorrow*. He saw the potential of smart cities as boundless, and transformative for cities in all countries. Energy production and distribution grids, health, transport, city government, and homes would all be 'smart'. The optimism of these three 'tech-fix' scenarios is perhaps even necessary, as we all need positive visions of the future to sustain us. What is important, but was not discussed in any of these possibilities for the future, was the supporting policies needed to bring such changes about, or any possible barriers to the application of big data to urban problems. The mere availability of big data was automatically assumed to lead to a better future for all.

This technological optimism is not new in the history of computers, or for general purpose technologies (e.g. steam engines, electricity, internal combustion engines, telephones) in general. The advent of personal computers in the 1970s, then the internet in the 1990s, and more recently, the new bio- and nano-technologies were similarly forecast to produce profound improvements, both to our lives and to the environment [30]. Instead, we have witnessed globally rising energy use and greenhouse gas emissions, species loss, and continued depletion of natural resources such as forests, fresh water, and minerals, including fossil fuels [29]. Nevertheless, it is the case that these general purpose technologies did profoundly change the way we live, even if not in the forecast direction. Big data is not exempt from this boosting of new technologies, as formalised in the Gartner Hype Cycle [9]. We need to repeat: we believe big data is a necessary but not sufficient condition for liveable, sustainable cities in the coming decades. Without the necessary policy changes, application of big data in cities would at best be ineffective, at worst it would leave us worse off in terms of urban sustainability from a global viewpoint and urban equity and health than at present.

How could big data actually be used in OECD cities, such as our own, Melbourne Australia, a city of 4.5 million, still growing mainly through international migration? First, we assume that the remaining technical problems facing use of big data earlier in the century can be overcome. Appreciation of both the many existing uses of big data and its potential benefits to all in a number of areas is now widespread: governments around the world are increasingly aware of its potential and are already devoting substantial resources to ensure that their country can gain the benefits from its application [15].

At the same time, it is possible that there will be a backlash against the use of personal data from ISPs and social media sites for commercial purposes, and that people will become far more careful about their postings on social media sites, as discussed above. This could lead to a reduction in big data useful for online marketing. On the other hand, the general public will be favourably disposed toward smart cities and the IoT to the extent that they see tangible benefits for themselves and the public good—and the extent to which they are actively involved in the decision-making process. Engaging the local community and stakeholders in the planning process will not only help provide a comprehensive understanding of the context but also help them to recognise the options and constraints in designing more liveable and sustainable cities. Such an outcome will involve, we argue, major changes to the existing smart city packages, which are still based on the assumption of endless economic growth, and the efficacy of technical fixes to solve any problems that arise.

Nevertheless, all this change will not happen without supporting policies at all levels of government, quite apart from growing government support for smart cities and other big data applications. First of all, the general changes outlined in Sect. 8.1 of this chapter will be required if civilised society has any chance to continue. Broad changes along these lines seem likely, assuming the human species and their leaders do want a future. Furthermore, specific changes will be needed, which may vary from city to city, depending on such factors as climate, per capita income, geography, national culture, and city size, among others. However, some general policies will apply to all cities. Some form of global carbon pricing will be needed [4], as discussed in Sect. 5.1. One possibility is that the tax is collected by an international agency, with a share of revenues redistributed to lower-income countries. A related point is the need to internalise other externalities, such as various forms of urban pollution so that the true cost of pollution is reflected in the higher prices people pay for fossil fuels and vehicular transport. Strong official policies need to be in place to tackle air pollution in all the cities of the world—it is no use having a plethora of data from sensors throughout the city if no effective action is taken.

Nevertheless, we will never attain zero pollution levels in our large cities, as control becomes progressively more difficult at higher levels of pollution removal. People also differ greatly in their sensitivity to air pollution, including natural pollutants like pollen. By providing data on the levels of particular pollutants (including forecasts for a few days ahead, as with weather forecasts), it will be possible for residents to take action to mitigate their deleterious health effects. The argument here is similar to that for dealing with climate change: mitigation is the best policy, but we also need a fall back policy to deal with the already apparent adverse effects of climate change. We need adaptation policies as well.

Increasingly, cities need to be viewed as a *system*, albeit a complex evolving one with multiple sustainability objectives, which can conflict with each other [34]. For example, high-density apartments in cities can save on winter heating energy, and reduce transport needs and energy use by encouraging non-motorised and public transport modes, while the inevitable road congestion discourages private transport. On the other hand, the UHI effect is more severe in densely populated cities, and it is more difficult to implement passive solar energy in multi-storey apartment blocks. Also, roof space per occupant will be less, which reduces the potential for PV solar panels and flat plate solar water heaters on rooftops to supply a significant fraction of household energy needs.

8.2.2   Big Data in the Cities of Industrialising Countries

Cities in the industrialising world differ as much from each other as they do from OECD cities—in their size, climate, transport systems, energy use and even income per capita. At least at the upper-income end, their problems, such as traffic congestion, are similar to those of OECD cities, and similar solutions should be more relevant.

In the OECD, the proportion of the population living in cities is already high, and population growth is low or even zero; hence the urban form of cities will change only slowly. In the rapidly urbanising parts of the world, however, city populations are growing rapidly. In China, new cities are being built from scratch, which presents the opportunity of designing cities with sustainability in mind. This is the intended aim of 'eco-cities'. One is presently being built at Masdar near Abu Dhabi in the United Arab Emirates, and another is planned: Dongtan City, to be sited on an island near Shanghai [35]. Experience to date shows that these cities will not only prove expensive but will take decades to build; both these drawbacks will limit their usefulness. Also, the plans rely heavily on technical fixes to achieve sustainability—they assume a business-as-usual world with continued economic growth.

The cities of presently very low-income countries, particularly in tropical Africa, face very different sustainability challenges. Very basic human needs are not met for a significant percentage of the population, especially those in unplanned settlements (see Chap. 2). Many urban residents do not have access to non-polluted water, adequate diet, housing, basic health provision or education. These needs should get priority; the question is whether big data applications such as the smart city concept can help achieve these aims in such cities? The answer, we think, is that they can, in some areas, but in other areas, at least for some time to come, they could prove a distraction, or worse, a hindrance. Like CCS and geoengineering, for some applications, proposed big data solutions are simply tech fixes promoted to avoid the need for deep social and political change.

Many cities in the poor countries of tropical Africa are growing in population but not in job provision. Services and infrastructure are inadequate even for existing residents, let alone rural migrants. According to World Bank data [44], in a number of countries, mainly in Africa, the proportion of urban residents with access to clean water is actually falling.

Paradoxically, where big data can help in these cities is in providing opportunities in *rural* areas to stem the flow to cities. For the foreseeable future, this means improvement of agricultural livelihoods [16, 17], or at the very least, averting famine in rural areas. Can the application of big data help here? On-going climate change and further rapid growth anticipated by the UN [41] are likely to constrain growth in food production per person. But national food production is only part of the problem; the other is the availability of food in local areas. Traditional information about food and nutrition security is at present often unsatisfactory. The use of big data (such as analysis of local social media messages about rising prices for staples) shows promise as a means of providing more timely and relevant information [24]. As another example of how mobile phone networks can potentially help agriculture, researchers have found that they 'could estimate how much precipitation was falling in an area by comparing changes in the signal strength between communication towers' [39]. This development could lead to improved precipitation forecasts at the very local scale.

As shown in Table 1.1, present electricity use per capita varies by nearly three orders of magnitude between nations, with lower but still very large, differences in primary energy use, and ownership levels of domestic appliances and cars. Although many countries, particularly those of tropical Africa, need higher per capita energy levels to ensure that the basic human needs of living already listed are met, attempting to follow the high energy consumption patterns of the Western countries—itself a product of the fossil fuel age [28]—is no longer a feasible option, for both resource and environmental reasons. They will need to follow a new path, just as they are now doing for phones. Land line ownership was always small in these countries, but mobile phone ownership is growing very rapidly, even in very low-income countries. They are already extensively used for information on weather and crop prices, and for small money transactions.

The already low energy use and travel levels of low-income cities imply that the more creative ways the residents use to cope with these low consumption levels will often hold valuable lessons for OECD cities. Conversely, there *are* lessons these low-income cities can learn from OECD cities—but trying to imitate their high consumption levels is not among them. Simple human solidarity demands the need for much convergence of consumption levels between presently low- and high-income cities. Just as they have bypassed land line telephony, they will also have to bypass private car travel as the dominant mode. Even so, traffic congestion in Chinese cities, and cities like Dhaka, Lagos or Bangkok is often even a worse problem than in OECD cities, and big data solutions might help here as they transition to lower traffic levels. Chapter 5 presented a possible PTA for Beijing, but similar systems would find application in OECD cities. An app for public transport information should help travellers in all cities.

Public health in low-income cities can already benefit from data mining. Tropical countries are already home to a variety of diseases not found in more temperate climates, and novel diseases can be expected to appear (Sect. 2.5). A key function of big data will be to identify outbreaks of epidemics and project their progress to better inform public health responses. Data from a range of sources, including satellite images, social media, travel (including air travel), and visits to local clinics can be mined for clues.

8.3   Discussion

The challenge facing cities, and the world in general, is simply this: how can we create the conditions in which all the world's people can have their basic needs met—needs that include not only obvious ones like food, fresh water and shelter, but also education, good health (both mental and physical) and cultural needs—and at the same time keep within the limits imposed by a planet with both finite resources and finite pollution absorption capability. At present, only for a minority of the global population are all these basic needs adequately met, yet as we have seen, the global Ecological Footprint is estimated at 1.5 Earths. In brief, we are far from meeting all basic needs, but even so, we are living unsustainably. To put the challenges we face in perspective, consider that if the world population in 2050—projected to be 9.725 billion by the UN [41]—each used 291 GJ per capita, the 2014 US primary energy use per capita, total global primary energy use would be 2830 EJ, or almost five times the actual 2014 global energy consumption [14]. And a number of countries today have much higher per capita energy consumption than the US.

The difficulty in achieving this goal, broadly similar to the UN Millennium Development Goals [40], will be made even more difficult with the expected continued growth in global population, with no peak expected before the end of this century [41], and with the decline in vital ecosystem services provided by the environment that such an increase in population, together with ongoing climate and environmental change, will bring. For cities, the difficulties can be compounded because available solutions can often conflict with each other, and further, solutions for one group in a given city can be at the expense of other groups.

The 2050 future outlined in the preceding sections is, of course, only one of several possible futures, some of them much bleaker than the optimistic ones given here by both the present authors and the other writers discussed. It may be that in addition to other problems besetting the planet, that IT itself will simply prove 'too fragile for the 21st century' [26], that security concerns will render the general use of internet-connected devices too risky. This book is evidently premised on the assumption that such is not the case, that although security problems will always be with us, the risks are manageable.

Although we have to predict the future and are often successful with trends that change only slowly, predicting the future path for Information Technology is even more difficult than in other areas, as the multitude of failed forecasts in this area testifies. Several key innovations—for instance, the impact of internet and the rise of smart phones—were either not predicted or their impact was greatly under-estimated. Or, as Vinton Cerf [7] put it in 1997: 'One of the more difficult challenges in trying to predict the future of technology is to distinguish between what will be commonplace from what may be merely feasible.' Hence the scenario offered here is partly normative, discussing how the big data revolution *could* eventually help tackle the many problems the world's cities face today.

But once again we stress that the potential for big data to help us move toward environmentally sustainable and healthy cities will usually come to nothing unless supporting policies are in place. First, the inequalities that are such an outstanding feature of today's world [44] must be drastically reduced, both between nations, and within nations and cities. *International* inequality is being reduced, but *intranational* inequality is worsening. A starting point here would be the fulfilment of the Millennium Development Goals of the UN. Chapter 2 stressed the pernicious effects of income and social inequality on health in all countries, rich and poor. In many low-income countries, these effects are further exacerbated by the unhealthy living conditions of slums, with their crowded conditions, poor sanitation, and their increased risk of infectious diseases—and for many, malnutrition. Big data can help in tracking and mapping inequality, but will not on its own help solve it. We believe there is a good chance that even failing simple concern for other members of the human family, enlightened self-interest in an increasingly interconnected world will prompt action on inequality.

The second area where strong policies are needed are those for combatting adverse climate and other environmental changes. Again, big data can (and increasingly is) being used to track relevant global and local environmental parameters. It can also contribute to system efficiency, helping us squeeze more services from a given level of energy inputs and pollution emissions. For decades, attention has been concentrated on improving device energy efficiency; the humans who use these devices were largely ignored. With the vast amounts of data becoming increasingly available on how we actually use household appliances and vehicles, this deficiency can now be addressed. But in the end, social changes which acknowledge that technical fixes can only take us so far for sustainability, are urgently needed.

Finally, the technical, economic, and social problems that big data will increasingly encounter must be solved (see Chap. 4). Although the meaning and scope of privacy are likely to change in an era of big data, it will still be important to people, and without its adequate protection, the big data future for urban sustainability will not happen, because of public opposition. Technical problems, broadly interpreted, here include not only the need to process, store, and finally make sense of very large amounts of data which may come from very different sources—for example, numerical data on temperature from environmental sensors and messages and even pictures on social media—but also security and reliability challenges. These various problems will affect the different aspects of urban sustainability differently, with some applications already presenting few of the challenges mentioned. It is from the initial application in these areas that the use of big data for urban sustainability can gain a foothold.

References

1. Airbus (2016) Global market forecast: mapping demand 2016–2035. http://www.airbus.com/company/market/global-market-forecast-2016-2035/. Accessed 10 Apr 2017. (Also earlier reports)
2. Anon (2017) Blocked by Google. New Scientist 234:7
3. Auliciems A (2009) Human adaptation within a paradigm of climatic determinism and change, Chapter 11. In: Ebi KL et al (eds) Biometeorology for adaptation to climate variability and change. Springer Science + Business Media BV, New York
4. Baranzini A, van den Bergh J, Carattini S et al (2015) Seven reasons to use carbon pricing in climate policy. http://www.uab.cat/web/research/working-papers-1293090813440.html. Accessed 10 Mar 2017
5. Bonney R, Shirk JL, Phillips TB (2014) Next steps for citizen science. Science 343:1436–1437
6. Budde P (2014) Smart cities of tomorrow, Chapter 12. In: Rassia ST, Pardalos PM (eds) Cities for smart environmental and energy futures, Energy systems. Springer, Berlin. https://doi.org/10.1007/978-3-642-37661-02
7. Cerf VG (1997) When they're everywhere. In: Denning P, Metcalfe R (eds) Beyond calculation: the next fifty years of computing. Springer, New York, pp 32–42
8. Ekanayake P, Moriarty P, Honnery D (2015) Equity and energy in global solutions to climate change. Energy Sustain Dev 26:72–78
9. Fisher S (2016) Gartner Hype Cycle 2016: blockchain, no big data. https://www.laserfiche.com/simplicity/gartner-hype-cycle-2016-blockchain-no-big-data/. Accessed 2 Mar 2017
10. Foley J (2017) Living by the lessons of the planet. Science 356:251–252
11. Hansen J, Sato M, Ruedy R (2012) Perception of climate change. Proc Natl Acad Sci U S A 109:E2415–E2423
12. Heinberg R (2010) Peak everything: waking up to the century of declines. New Society Publishers, Gabriola Island, BC
13. Intergovernmental Panel on Climate Change (IPCC) (2015) Climate change 2014: synthesis report. Cambridge University Press, Cambridge, UK
14. International Energy Agency (IEA) (2016) Key world energy statistics 2016. IEA/OECD, Paris
15. Jin X, Wah BW, Cheng X et al (2015) Significance and challenges of big data research. Big Data Res 2:59–64
16. Kshetri N (2014) The emerging role of Big Data in key development issues: opportunities, challenges, and concerns. Big Data Soc 1:1–20
17. Kshetri N (2016) Big data's big potential in developing economies: impact on agriculture, health and environmental security. CABI Press, Boston
18. Kubiszewski I, Costanza R, Franco C et al (2013) Beyond GDP: measuring and achieving global genuine progress. Ecol Econ 93:57–68

19. Le Bars D, Drijfhout S, de Vries H (2017) A high-end sea level rise probabilistic projection including rapid Antarctic ice sheet mass loss. Environ Res Lett 12:044013. https://doi.org/10.1088/1748–9326/aa6512
20. Le Page M (2017) CO_2 levels to hit 50-million-year high by 2050. New Scientist 234:9
21. Le Page M (2017) The price of emission. New Scientist 234:22–23
22. Marsh M, Coughlan A (2013) China's struggle to clear the air. New Scientist 217:8–9
23. Miller-Rushing A, Primack R, Bonney R (2012) The history of public participation in ecological research. Front Ecol Environ 10(6):285–290
24. Mock N, Morrow N, Papendieck A (2013) From complexity to food security decision-support: novel methods of assessment and their role in enhancing the timeliness and relevance of food and nutrition security information. Glob Food Sec 2:41–49
25. Montgomery B (2015) Future shock: IoT benefits beyond traffic and lighting energy optimization. IEEE Consum Electron Mag 4:98–100
26. Moriarty P (1999) Don't depend on IT. Aust Q 71(6):16–20. 40
27. Moriarty P (2016) Reducing levels of urban passenger travel. Int J Sustain Transp 10(8):712–719
28. Moriarty P, Honnery D (2011) Rise and fall of the carbon civilisation. Springer, London
29. Moriarty P, Honnery D (2014) The Earth we are creating. AIMS Energ 2(2):158–171. https://doi.org/10.3934/energy.2014.2.158
30. Moriarty P, Honnery D (2014) Reconnecting technological development with human welfare. Futures 55:32–40
31. Moriarty P, Honnery D (2015) Future cities in a warming world. Futures 66:45–53
32. Moriarty P, Honnery D (2015) Reliance on technical solutions to environmental problems: caution is needed. Environ Sci Technol 49:5255–5256
33. Moriarty P, Honnery D (2016) Can renewable energy power the future? Energy Policy 93:3–7
34. Moriarty P, Honnery D (2017) Creating environmentally sustainable cities: not an easy task. In: Archer K, Bezdecny K (eds) Handbook of emerging 21st century cities. Edward Elgar, London
35. Premalatha M, Tauseef SM, Abbasi T et al (2013) The promise and the performance of the world's first two zero carbon eco-cities. Renew Sust Energ Rev 25:660–669
36. Randers J (2013) Global trends 2030 compared with the 2052 global forecast. World Future Rev 5(4):360–366
37. Ravilious K (2013) Mid-town miners. New Scientist 218:40–43
38. Steffen W, Richardson K, Rockström J et al (2015) Planetary boundaries: guiding human development on a changing planet. Science 347(6223):1259855. (10pp)
39. Tollefson J (2017) Rain forecas go mobile. Nature 544:146147
40. United Nations (UN) (2008) The millennium development goals report 2008. UN, New York
41. United Nations (UN) (2015) World population prospects: the 2015 revision. http://esa.un.org/unpd/wpp/Download/Standard/Population/. Accessed 15 Dec 2016
42. van den Bergh JCJM (2017) A third option for climate policy within potential limits to growth. Nat Clim Chang 7:107–112
43. Williams M, Zalasiewicz J, Waters CN et al (2016) The Anthropocene: a conspicuous stratigraphical signal of anthropogenic changes in production and consumption across the biosphere. Earth's Future 4:34–53. https://doi.org/10.1002/2015EF000339
44. World Bank (2017) Indicators. http://data.worldbank.org/indicator. Accessed 28 Apr 2017

Index

A
Activity shifting, 144
Air pollution
    effect of climate change, 31, 33
    PM 2.5 levels, 25, 27, 33, 36, 37
Algorithms
    bias potential, 67
    definition, 50
Amazon, 56, 57, 75, 143
Anthropocene epoch, 142
Artificial intelligence (AI), 50, 58, 60,
        75, 122
Australia, 8, 45, 54, 85, 86, 89, 129, 149
Automated vehicle (AV)
    energy savings, 93, 116
    risks, 93
    safety, 148

B
Bangladesh, 33, 95, 96
Beijing, 17, 27, 36, 37, 53, 55, 85, 86, 95–98,
        100, 152
Big data
    characteristics, 90
    cost, 49, 74
    definitions, 48, 50
    energy requirements, 116
    measurement error, 50
    potential, 57, 81
    present use, 55
    privacy, 17, 60, 65, 74
    reliability, 17, 60, 65, 71, 72, 74, 154
    risks, 136

sampling error, 45
security, 60, 65, 69, 70, 74, 154
technical problems, 73–74, 76, 149, 154
Big data applications
    energy conservation, 106
    existing uses, 83, 149
    health, 69, 122–124
    predictive maintenance, 113
    transport, 54, 81–101
Bioenergy carbon capture and storage
        (BECCS), 4, 6
Boston, 52, 54, 57, 86
Brisbane, 86

C
Carbon dioxide (CO_2) removal, 4, 6
Carbon pricing, 91, 96, 106, 146, 149
China
    air pollution, 36, 37
    urban growth, 23, 34
Cities
    environmental sustainability, 2, 9,
        51, 146
    share of global emissions, 47
    vulnerability to natural hazards, 10
Climate change
    adaptation, 11, 145
    climate forcing, 6, 7, 10
    disease risk, 29
    sea level rise, 4, 10, 142
    urban health, 28
Climate forcing, 6, 7, 10
Closed-circuit TV (CCTV), 68, 82

D
Data needs, 46–48, 97, 127
Delhi, 27, 33, 34
Dementia patients, 123
Desertec project, 107
Dhaka, 10, 33, 95, 96, 152

E
Ebola outbreak, 126–128
Eco-cities, 150
Eco-efficiency, 12, 23, 97
Ecological overshoot, 142
Electric vehicles
    charging, 111, 112
    impact on grids, 111, 112
    vehicle-to-grid energy storage, 110
Energy, 2, 4
    combined heat and power (CHP), 15, 114
    domestic energy use, 12, 15, 90, 108, 111
    efficiency, 5–7, 9, 12–15, 17, 84, 91, 92,
        94, 95, 107, 153
    fossil (*see* Fossil fuels)
    passive cooling, 145
    perverse incentives, 109
    prosumers, 110
    renewable (*see* Renewable energy)
    storage, 5, 73, 108, 111, 112
    waste heat, 114
Environmental sustainability
    challenges, 3–16
    global, 2–7
    solutions, 3–16
    urban, 2, 8–16

F
Floating Car Data (FCD), 82, 83
Fossil fuels
    depletion, 7, 105, 149
    subsidies, 7, 88, 144, 146
Future
    car ownership, 101
    vehicular travel, 100
    year 2050 world, 100

G
Graphical Information System (GIS), 97
Greenhouse gas (GHG), 2, 4, 7, 10–12,
        14–16, 28, 36, 59, 84, 85, 87, 90,
        92–95, 111–114, 121, 141, 142, 144,
        146, 149
Gross National Income (GNI), 8, 26

H
Health
    application examples, 124, 128
    climate change effects, 28
    data needs, 97
    global, 23–26, 119, 126
    mental, 23, 26–27, 29, 30, 33, 123, 146
    rising costs, 55, 120
    urban, 18, 23–40, 48, 51, 55, 59, 87, 119,
        146, 147
Healthcare, *see* Health
Human Development Index (HDI), 26

I
Inequality
    international, 153
    intranational, 153
Intelligent cities, *see* Smart cities
Intelligent Highway Vehicle Systems (IHVS), 92
Intergovernmental Panel on Climate Change
        (IPCC), 4, 6, 7, 9, 11, 28, 105, 106, 145
Intermittent renewable energy, 9, 47, 105, 107,
        108, 110, 111
International Atomic Energy Agency (IAEA), 5
International Council for Local Environmental
        Initiatives (ICLEI), 2
Internet of Things (IoT)
    definition, 48–51
    growth, 59
    sensors, 17, 50, 52

L
Life cycle analysis (LCA), 15
Liveability, 2, 3, 30, 31, 33–39
Longevity, 23, 24, 26, 29, 40, 119

M
Machine learning, 50, 54, 75, 113, 127
Medicine, *see* Health

N
National Health System (UK) (NHS), 122
Natural experiments, 90
Negative externalities, 7, 88
Noise pollution, 12, 14, 27, 28, 52, 53, 85,
        121, 143, 147

O
Oil depletion, 47

P

Pedestrian traffic counts, 54
Personal Travel Assistant (PTA), 81, 87, 88, 90, 95, 96, 152
Planetary limits, 60, 143
Points of Interest (POI), 53, 81, 83
Privacy, *see* Big data
Problem shifting, 6

Q

Quality of life (QOL), 1, 2, 17, 26, 38, 51, 60, 75, 122, 146
Quantified self (QS)
    definition, 119, 124, 125
    present use, 119

R

Reliability, *see* Big data
Renewable energy
    intermittency problem, 106, 107
    solar, 5, 47
    wind, 5, 47, 107
Risk, *see* Big data

S

Security, *see* Big data
Self help groups, 124, 125
Smart appliances, 112, 144, 148
Smart buildings, 112, 113
Smart cities
    definition, 50
    examples, 51
Smart grids
    advantages, 106–108
    properties, 108, 110
Smart meters, 18, 54, 109, 122, 148
Smart phones, 18, 54, 73, 86, 88, 89, 122–124, 128, 143, 147, 153
Solar radiation management (SRM), 5–7, 10, 142
Sustainability, 1–18, 23, 38, 45–60, 65, 68, 70, 75, 77, 84, 85, 88–90, 94, 101, 105, 106, 112, 115, 116, 141, 146–150, 153, 154
Sustainable cities, 2, 16, 46–48, 74, 92, 114, 149

T

Technological optimism, 148
Tele-work, 89

Traffic
    congestion, 1, 9, 13, 28, 53, 54, 83, 86, 89, 96, 150, 152
    parking, 53, 54, 57
    reductions, 89–91
Transport
    efficiency, 12, 14
    freight, 12, 16, 91, 92
    monoculture, 87
    non-motorised, 28, 54, 84–88, 90, 93, 94, 101, 144, 150
    occupancy rates, 13, 85, 90, 95
    polyculture, 87
    public, 12, 13, 54, 55, 57, 68, 76, 85–88, 90, 92, 94, 95, 98, 100, 112, 144, 150, 152
    surveys, 45
    sustainability, 17, 88
    trajectory data, 82
    urban, 12–14, 46, 48, 81, 108, 147
Travel
    comfort, 81, 85, 94, 95
    convenience, 14, 90, 94
    discretionary, 90
    forecasts, 100, 101
    non-discretionary, 90, 91
    routines, 86, 91
    substitution with IT, 89
    wellbeing, 30, 94, 95

U

Uber, 59, 86
Urban
    data collection, 45, 46
    data needs, 46, 48
    density, 95, 114
    ecology, 2, 11
    environment, 2, 8–16, 38, 48, 58, 59, 81, 82, 98
    governance, 51, 55, 59
    health, 18, 23, 48, 51, 55, 59, 87, 119–136, 146, 147
    heat waves, 10
    infrastructure, 51, 52, 69, 115
    sea level rise, 10
Urban Heat Island (UHI)
    causes, 10, 11, 29
    definition, 10, 115
Urban Transport Energy Saver (UTES)
    description, 97, 98, 100
    travel simulation screen display, 97

Urbanisation
    China, 31, 34–37
    global, 16

V
Vehicles
    alternative fuels, 12
    efficiency, 12, 85, 94
Virtual spine, 18, 129–132, 135, 136

W
Waste management, 52, 54, 58
Wellbeing, 17, 18, 23, 51, 55, 57, 58, 94, 95,
        116, 119, 146, 147

Z
Zhuhai city, 38, 39

Printed by Printforce, the Netherlands